T0398216

Natural Heritage from East to West

Niki Evelpidou · Tomás de Figueiredo ·
Francesco Mauro · Vahap Tecim ·
Andreas Vassilopoulos
Editors

Natural Heritage from East to West

Case studies from 6 EU Countries

Editors

Dr. Niki Evelpidou
National and Kapodistrian
University of Athens
Faculty of Geology
and Geoenvironment
Department of Geography
and Climatology
Panepistimiopolis, Gr-15784
Athens, Greece
evelpidou@geol.uoa.gr

Dr. Tomás de Figueiredo
CIMO, Escola Superior Agrária
Instituto Politécnico de Bragança
E. S. Agrária
CIMO-Moutain Research Centre
Apartado 1172, 5301-855
Bragança, Portugal
tomasfig@ipb.pt

Dr. Francesco Mauro
Università Telematica
Guglielmo Marconi
Via Plinio 44, 00193
Rome
Italy
mauro_sustainability@yahoo.com

Dr. Vahap Tecim
Dokuz Eylül University
Faculty of Economics
and Administrative Science
35160, Buca/Izmir, Turkey
vahap.tecim@deu.edu.tr

Dr. Andreas Vassilopoulos
Geoenvironmental Institute
Flias 13, Maroussi
151 25, Athens, Greece
vassilopoulos@geoenvi.org

ISBN 978-3-642-01576-2 e-ISBN 978-3-642-01577-9
DOI 10.1007/978-3-642-01577-9
Springer Heidelberg Dordrecht London New York

Library of Congress Control Number: 2009932879

Cover design: Bauer, Thomas

The authors, in cooperation with their national coordinator (quoted below), take full responsibility on the originality and the copyrights of the provided material:
Greece: Evelpidou N., (National and Kapodistrian University of Athens, Faculty of Geology and Geoenvironment, Department of Geography and Climatology, Panepistimiopolis, 15784, Athens, Greece), Vassilopoulos A., (Geoenvironmental Institute, Flias 13, Maroussi, 151 25, Athens, Greece),
Italy: Mauro F. (Università Telematica "Guglielmo Marconi", Via Plinio 44, 00193 Rome, Italy),
Malta: Integrated Resources Management Company Ltd., Malta,
Portugal: de Figueiredo T. (Instituto Politécnico de Bragança, E. S. Agrária, CIMO-Moutain Research Centre, Bragança, Portugal),
Romania: Ioane D., Marunteanu C. (University of Bucharest, Romania),
Turkey: Tecim V. (Dokuz Eylul University, Faculty of Economics and Administrative Sciences, 35160 Buca, Izmir-Turkey)
Printed by SPRINGER

Education and Culture

Leonardo da Vinci e-duNatHer training web tools 4 everyone

The project is co-funded by the European community (75%)
The book doesn't necessarily reflect the views of the European Commission or the National Agency

Printed on acid-free paper

Springer is part of Springer Science+Business Media (www.springer.com)

Foreword

Cumulative global transformations, occurring daily, affect important aspects of our life. Characteristic cultural and natural heritage, including sites of priceless value, is under constant threat. There are growing pressures, of both natural and human origin, such as wars, conflicts, natural or technological disasters and the effects of global climate change. These provoke the continuous degradation of many sites included in the World Heritage List. In consequence, immediate strategic measures must be taken.

Natural heritage is our legacy from the past, that we inherited from our ancestors and pass on to future generations. It is vital to realize its value and protect it by all possible means, enforcing innovative and sustainable action plans that promote global international co-operation.

This book aims to address specific natural heritage sites in Europe, from West to East. The six countries of study interest are Portugal, Malta, Greece, Italy, Romania and Turkey. For each case, the corresponding current status is presented. This is accompanied by recommended action plans for protection and conservation, training initiatives that improve the public awareness of natural heritage issues and efforts to estimate the natural/environmental value of the sites. The book is the overall result of an interregional initiative aiming to promote convergence, provoke public interest and recommend action for radical changes in our attitude towards heritage conservation.

Athens, Greece	Andreas Vassilopoulos
Athens, Greece	Niki Evelpidou
Bragança, Portugal	Tomás de Figueiredo
Rome, Italy	Francesco Mauro
Buca/Izmir, Turkey	Vahap Tecim
2009	

Acknowledgements

GeoEnvironmental Institute would like to thank all editors and authors that contributed to this book. Also, a special acknowledgment must be extended to MSc Konstantia Chartidou (School of Geology and GeoEnvironment, University of Athens) for her support to this project and the proofing of the content and to Dr. John Peterson (School of Computer Sciences, University of East Anglia) for corrections and improvements to English text.

All authors would like to acknowledge the contribution of Dr. John W.M. Peterson, School of Computing Sciences, University of East Anglia, Norwich, UK, for corrections and improvement to the English text.

Università Telematica Guglielmo Marconi (UTGM) and ENEA acknowledge the collaboration of CUTGANA (Centro Universitario per la Tutela e la Gestione degli Ambienti Naturali e degli Agroecosistemi), Università di Catania, for the paper on "The Cyclops Islands".

IRMCo acknowledges the use of the Integrated Land and Water Information System (ILWIS), developed by ITC, the Netherlands, for the management and assessment of geographic information in a GIS environment. ILWIS functionality was employed for the paper on "The natural heritage of the Island of Gozo" and the paper on "The geomorphological cave features of Għar il-Friefet".

IPB (Polytechnic Institute of Bragança) wishes to acknowledges all those colleagues, most of them also members of CIMO (Centre for Mountain Research), that contributed to the recently issued Management Plan of Montesinho Natural Park (PNM). Their hidden contribution to the articles concerning PNM is much acknowledged. A word in recognition of his endless and contagious enthusiasm towards Montesinho and to the Mountain domain, spread among us all in the IPB, is due to Professor Dionísio Gonçalves, the first Director, Coordinator and President of PNM, CIMO and IPB, respectively. The authors of the photos inserted in the articles concerning Montesinho are also much acknowledged for their contribution, which is referenced at the end of each text, unless they are authors of papers, too.

Contents

Contributors

Calos Aguiar Instituto Politécnico de Bragança, Escola Superior Agrária, Mountain Research Center – CIMO, Campus Sta Apolónia, 5301-855 Bragança, Portugal, cfaguiar@ipb.pt

Ebru Alakavuk Yasar University Kazim Dirik Mahallesi 364 Sok, No:5, Bornova, İzmir, Turkey, ebru.alakavuk@yasar.edu.tr

Cláudia Araújo CIBIO – Centro de Investigação em Biodiversidade e Recursos Genéticos; Departamento de Botânica, Faculdade de Ciências da Universidade do Porto, Portugal, claudia.araujo@fc.up.pt

Theodoros Astaras Department of Physical and Environmental Geography, School of Geology, Aristotle University of Thessaloniki, University Campus 541 24 Thessaloniki-Greece, astaras@geo.auth.gr

Ebru Aydeniz Yasar University Kazim Dirik Mahallesi 364 Sokak No: 5, Bornova, İzmir, Turkey, ebru.aydeniz@yasar.edu.tr

Muhammed Aydoğan Faculty of Architectural, Dokuz Eylul University, City and Regional Planning Department & Department of GIS, m.aydogan@deu.edu.tr

Gülnur Ballice Yasar University İzmir, 35500, Turkey, gulnur.ballice@yasar.edu.tr

Sabah Balta Department of Tourism and Hotel Management, Yasar University, Izmir, Turkey, sabah.balta@yasar.edu.tr

Horea Bedelean "Babes-Bolyai" University, Cluj-Napoca, Romania, bedelean@bioge.ubbcluj.ro

João Bento Universidade de Trás-os-Montes e Alto Douro, Departamento Florestal, Vila Real, Portugal, j_bento@utad.pt

Paola Carrabba Department of Biotechnology, Agro-industry and Health Protection, ENEA – National Agency for New Technology, Energy and the Environment, Casaccia Research Center, 00123 Rome, Italy, carrabba@casaccia.enea.it

J. Castro Instituto Politécnico de Bragança, Escola Superior Agrária Apartado 1172, 5301-855 Bragança, Portugal, mzecast@ipb.pt

João Paulo Castro Instituto Politécnico de Bragança, Escola Superior Agrária, Mountain Research Centre – CIMO, Campus Sta Apolónia, 5301-855 Bragança, Portugal, jpmc@ipb.pt

Yvonne Cerqueira CIBIO – Centro de Investigação em Biodiversidade e Recursos Genéticos; Departamento de Botânica, Faculdade de Ciências da Universidade do Porto, Portugal, yvonne.cerqueira@fc.up.pt

Konstantia Chartidou National and Kapodistrian University of Athens, Faculty of Geology and Geoenvironment, Department of Geogrpahy and Climatology, Panepistimiopolis, Gr-157 84, Athens, Greece, chartidou@geoenvi.org

Celeste Oliveira Alves Coelho Department of Environment and Planning, CESAM – Centre for Environmental and Marine Studies, University of Aveiro, Portugal, coelho@dao.ua.pt

Osman Culha Department of Tourism and Hotel Management, Yasar University Kazim Dirik M., 364 S. No: 5, 35500, Bornova, Izmir, Turkey, osman.culha@yasar.edu.tr

Tomás de Figueiredo Instituto Politécnico de Bragança, Escola Superior Agrária, Mountain Research Centre – CIMO, Campus Sta Apolónia, 5301-855 Bragança, Portugal, tomasfig@ipb.pt

Dirk De Ketelaere Integrated Resources Management (IRM) Co. Ltd. Malta, info@environmentalmalta.com

Niki Evelpidou National and Kapodistrian University of Athens, Faculty of Geology and Geoenvironment, Department of Geography and Climatology, Panepistimiopolis, Gr-15784, Athens, Greece, evelpidou@geol.uoa.gr

Pedro Ferreira Universidade de Trás-os-Montes e Alto Douro, Departamento Florestal, Vila Real, Portugal, pedrof@utad.pt

Felícia Fonseca Instituto Politécnico de Bragança, Escola Superior Agrária, Mountain Research Centre – CIMO, Campus Sta Apolónia, 5301-855 Bragança, Portugal, ffonseca@ipb.pt

Jesús Garrido Integrated Resources Management (IRM) Co. Ltd. Malta, info@environmalta.com

João Honrado CIBIO – Centro de Investigação em Biodiversidade e Recursos Genéticos; Departamento de Botânica, Faculdade de Ciências da Universidade do Porto, Portugal, jhonrado@bot.fc.up.pt

Dumitru Ioane University of Bucharest, Romania, dumitru.ioane@g.unibuc.ro

Athanassios Katerinopoulos National and Kapodistrian University of Athens, Department of Geology and Geoenvironment, Section of Mineralogy and Petrology, Panepistimiopolis, Gr-15784, Athens, Greece, akatern@geol.uoa.gr

Akindinos Kelepertzis National and Kapodistrian University of Athens, Faculty of Geology and Geoenvironment, Department of Economic Geology and Geochemistry, Greece, kelepertsis@geol.uoa.gr

Konstantinos G. Kyriakopoulos National and Kapodistrian University of Athens, Department of Geology and Geoenvironment, Section of Mineralogy and Petrology, Panepistimioupolis, Gr-15784, Athens, Greece, ckiriako@geol.uoa.gr

Dimitra Leonidopoulou National and Kapodistrian University of Athens, Faculty of Geology and Geoenvironment, Depatment of Geography and Climatology, Panepistimiopolis, Gr-157 84, Athens, Greece, dleonid@geol.uoa.gr

Domingos Lopes Universidade de Trás-os-Montes e Alto Douro, Departamento Florestal, Vila Real, Portugal, dlopes@utad.pt

Marco Magalhães Universidade de Trás-os-Montes e Alto Douro, Departamento Florestal, Vila Real, Portugal, mpmmaga@utad.pt

Jorge Espinha Marques Universidade do Porto, Faculdade de Ciências, Departamento de Geologia/Centro de Geologia, Porto, Portugal, jespinha@fc.up.pt

José Manuel Marques Universidade Técnica de Lisboa, Instituto Superior Técnico, Departamento de Engenharia de Minas e Georrecursos, Lisboa, Portugal, jose.marques@ist.utl.pt

Cristian Marunteanu University of Bucharest, Romania, crimarunteanu@yahoo.com

Francesco Mauro Università Telematica "Guglielmo Marconi", Via Plinio 44, 00193 Rome, Italy, mauro_sustainability@yahoo.com

Carlos Meireles Instituto Nacional de Engenharia, Tecnologia e Inovação, Departamento de Geologia, S. Mamede de Infesta, Portugal, Carlos.Meireles@ineti.pt

Antonela Neacsu University of Bucharest, Romania, antonela@geo.edu.ro

Sílvia Nobre Instituto Politécnico de Bragança, Escola Superior Agrária, Mountain Research Centre – CIMO, Campus Sta Apolónia, 5301-855 Bragança, Portugal, silvian@ipb.pt

Gokce Ozdemir Department of Tourism and Hotel Management, Yasar University Kazim Dirik M., 364 S. No: 5, 35500, Bornova, Izmir, Turkey, gokce.ozdemir@yasar.edu.tr

Malike Özsoy Yasar University, 35500 Bornova-İZMİR, malike.ozsoy@yasar.edu.tr

Laura Maria Padovani Department of Biotechnology, Agro-industry and Health Protection, ENEA – National Agency for New Technology, Energy and the Environment, Casaccia Research Center, 00123 Rome, Italy, padovani@casaccia.enea.it

Henrique Miguel Pereira CBA – Centro de Biologia Ambiental, Faculdade de Ciências da Universidade de Lisboa; ICNB – Instituto da Conservação da Natureza e da Biodiversidade, Portugal, hpereira@fc.ul.pt

Luís Carlos Pires Instituto Politécnico de Bragança, Escola Superior Agrária, Mountain Research Centre – CIMO, Campus Sta Apolónia, 5301-855 Bragança, Portugal, luica@ipb.pt

Alessandro Ramazzotti Università Telematica "Guglielmo Marconi", Via Plinio 44, 00193 Rome, Italy, alessandro.ramazzotti@macchind.it

Ilaria Reggiani Università Telematica "Guglielmo Marconi", Via Plinio 44, 00193 Rome, Italy, progetti16@unimarconi.it

Ferika Özer Sari Yasar University, 35500 Bornova-İZMİR, ferika.ozersari@yasar.edu.tr

Burcu Şengün Yasar University Kazim Dirik Mahallesi 364 Sok, No:5 Bornova, İzmir, Turkey, burcu.sengun@yasar.edu.tr

Eugénio Sequeira Liga para a Protecção da Natureza, Lisboa, Portugal, eugenio.sequeira@sapo.pt

Michael Sinreich Integrated Resources Management (IRM) Co. Ltd. Malta

Anna Spiteri Integrated Resources Management (IRM) Co. Ltd. Malta, info@environmalta.com

Alexandru Szakács Sapientia University, Cluj-Napoca; Institute of Geodynamics, Romanian Academy, Bucharest, Romania, szakacs@sapientia.ro

Evagelos Tziritis National and Kapodistrian University of Athens, Faculty of Geology and Geoenvironment, Department of Economic Geology and Geochemistry, Greece, evtziritis@geol.uoa.gr

Andreas Vassilopoulos Geoenvironmental Institute, Flias 13, Maroussi, 151 25, Athens, Greece, vassilopoulos@geoenvi.org

Josianne Vella Integrated Resources Management (IRM) Co. Ltd. Malta, info@environmalta.com

Joana Vicente CIBIO – Centro de Investigação em Biodiversidade e Recursos Genéticos; Departamento de Botânica, Faculdade de Ciências da Universidade do Porto, Portugal, jsvicente@fc.up.pt

Zeynep Yağmuroğlu Yasar University Kazim Dirik Mahallesi 364 Sok, No:5 Bornova, İzmir, Turkey, zeynep.yagmuroglu@yasar.edu.tr

Nickolas C. Zouros Department of Geography, University of the Aegean, GR–81100; Natural History Museum of the Lesvos Petrified Forest, Lesvos, GR–81112, Greece, nzour@aegean.gr

Sandra Valente Department of Environment and Planning, CESAM – Centre for Environmental and Marine Studies, University of Aveiro, Portugal, sandra.valente@ua.pt

Cristina Ribeiro Department of Environment and Planning, CESAM – Centre for Environmental and Marine Studies, University of Aveiro, Portugal, cristinaribeiro@ua.pt

Geomorphological Evolution of Santorini

Andreas Vassilopoulos, Niki Evelpidou, and Konstantia Chartidou

Santorini is an island of the Aegean Sea that belongs to the Prefecture of Cyclades. It is located southern of Ios Island and, along with Anafi, these are the southernmost islands of the Cyclades. Santorini is composed of Thera, with a crescent shape, and the islands of Therassia and Aspro (Aspronisi) in a circle. In the centre of the circle lies the caldera, which was formed by a volcanic eruption (or eruptions) and the simultaneous collapse of a part of the island. Santorini caldera is one biggest of the world, covering an area of approx. 83 km^2, with a length of 11 km (N–S) and a width of 7.5 km (E–W). The volcanic islands of Nea Kameni and Palaia Kameni have formed within the caldera. Nowadays Santorini is a volcanic island that belongs to the Aegean volcanic arc and, with its fumaroles, gases and a high temperature, is the only active volcano in the Eastern Mediterranean.

Santorini complex belongs to Cyclades islands situated into Aegean Sea

A. Vassilopoulos (✉)
Geoenvironmental Institute, Flias 13, Maroussi, 151 25, Athens, Greece
e-mail: vassilopoulos@geoenvi.org

N. Evelpidou et al. (eds.), *Natural Heritage from East to West*,
DOI 10.1007/978-3-642-01577-9_1, © Springer-Verlag Berlin Heidelberg 2010

The conceivable circle forming the caldera of Santorini. This circle is composed by Thera, Therasia and Aspro islands

Parts of the caldera walls in Thera island

Santorini has been the subject of numerous studies undertaken by a wide variety of scientists (Lacroix, 1896; Padang, 1936; Marinatos, 1939, 1968; Galanopoulos, 1958 & 1971; Ninkovich et al. 1965; Heinek-Mac-Coy, 1984; Velitzelos, 1990; Skarpelis and Liati, 1987; Lagios et al. 1989; Fytikas et al. 1990). Nevertheless, geomorphological publications are still rare. Stratigraphically, the lithographic complex of Santorini consists of two main categories of rocks. At the base of the complex the following formations are found: Triassic crystalline limestones and dolomites, Eocene phyllites and a Miocene granite intrusion. On top of these, a continuous sequence of volcanic rocks dominates the island; it began to form approximately 1.6 million years ago and continued up to modern times.

a) b) c)

Within the caldera the volcanic islands of Nea Kameni and Palaia Kameni have been formed

Hot springs in Thera island

Santorini's volcanic sequence is characterised by an alternation of volcanic lavas and pumice. Literature often distinguishes three stratigraphic horizons of pumice, and refers to them as, lower, medium and superior pumice horizons. It should be noted that the superior pumice horizon is the result of the Minoan explosion, dating to around 1,500 BC.

The explosions of Santorini volcano began at the end of the Neocene period, about 2 million years ago, after Aigiida broke into pieces and sunk. The island that existed before the volcanic eruptions was residue of a crystalloschistosive mass. Residues of that pre-volcanic island are found in the Profitis Ilias area, which today has the higher altitude in the island (568 m). The most ancient volcanic centres are located on the southern part of the island. The historical eruptions of this volcano were first recorded in 197 BC (Strabo).

Volcanic lavas in Thera island a) in the caldera walls b) in red beach

- Due to the explosion in 197–199 BC a small islet – Iera – appeared between Thera and Therasia.
- During the 19 AD explosion an islet named Thera was formed.
- With a powerful explosion in 46 AD, which lasted a few months, another islet near Iera was created.
- In 726 AD a violent explosion probably caused damages to agriculture due to the ash fall. The same explosion resulted in the merging of the Iera islet with the one created in 46 AD. None of these islets exists today and they are in no way relevant to the islands of Kameni.
- During the 1570–1573 explosion Palaia Kameni emerged 66 m above sea level.
- In 1650 an explosion took place 6.5 Km outside the caldera, eastern of Cape Columbo, and a small island formed. This island later sunk, leaving in its place Columbo reef, 19 m below the sea surface.
- The explosions of the 1707–1711 period created Nea Kameni.
- During the 1866–1870 explosions the "George I" (in honour of King Geoge I) dome was created, 130.8 m high and constituting the highest altitude in modern Nea Kameni, accompanied by the Afroessa volcanic cone.

Profitis Ilias mountain has the higher altitude in the island and is a prevolcanic relict consisting mainly of crystallised limestones

- The 1925 (August 11th) and 1926 (May 31st) explosions, created the Dafni dome (in honour of the cruiser who visited the island), that merged Palaia Kameni and Nea Kameni, which was already merged to Afroessa and George I.
- On January 23rd and March 17th 1928, a new explosion created the Nautilus dome (named after the ship of the Greek Hydrographical Service).
- From 1939 to 1941 new explosions in the area of George I and Nea Kameni created the domes Triton, Ktenas, Fouqué, Smith, Reck and Niki (named in honour of Greece's victory in the Greek-Italian war).
- In 1950 the volcano's reactivation resulted in the formation of Liatsikas dome (named after the Greek geologist Liatsikas who studied the Santorini volcano).

a) b)

Part of the Minoan volcanic relief is smooth and has been formed by tuffs deposition

Given the complexity of the volcanic evolution, various statistical methods have been used (Vassilopoulos et al., 2002) to analyse the diverse geomorphological units and various software programs were applied in order to visualise the alteration of the relief.

Parts of the caldera which was mainly formed during the Minoan eruption

Five geomorphological units have been distinguished in Thera Island:

- The limestone–schist Unit (Prophitis Ilias – Vlychada). This formation is found in the south–eastern part of the island and mainly consists of crystalic limestones, dolomites and phyllites representing the pre-volcanic relief of Thera.
- The volcanic clusters (Mesa Vouno – Mikros Prophitis Ilias – Skarou) located in the Northern part of the island, mainly formed by andesitic lavas.
- The Minoan volcanic relief of Thera and Therassia shapes in the form of a crescent. This unit can be further divided in two sub-units, the one having a smooth relief and the other a rough one.
- The Caldera is the most impressive landform in Thera and has attracted many researchers over time.
- The Unit of the newer volcanic islands, Palaia and Nea Kammeni, whose creation expresses the most recent volcanic activity.

a)

b)

The newer volcanic islands, Palaia and Nea Kammeni, whose creation express the most recent volcanic activity. Figure a) has been taken from Nea Kammeni and shows Palaia Kammeni. Figure b) Show the volcanic crater of Nea Kammeni during a geological field work

The Akrotiri area is believed to have been inhabited at least since the late Neolithic period (7,000 BC–3,500 BC). Akrotiri settlement is dated by ceramics to as far back as the mid 4th millennium BC. These ceramics are related to those of the Neolithic settlement of Saliagos, between Paros and Antiparos. Almost 50 years of excavations have brought to light a self-sufficient settlement with a network of water supply and sewerage, along with paved roads and houses.

Ceramic elements found in Akrotiri settlement, dated at 4th millennium BC

A very well sewerage system was developed in Akrotiri village

Thick volcanic material deposits covered the prehistoric village of Akrotiri and conserved two or even three floor buildings, with their whole contents intact. The excavated section is a very small part of the entire settlement. Akrotiri must have covered an area around 200,000 m^2. Nine buildings are known but none has yet been entirely studied. The southernmost part of the excavation is 250 m away from the current beach. The settlement spreads along the slopes of a small hill, it is arranged like an amphitheatre with slopes towards the S and E.

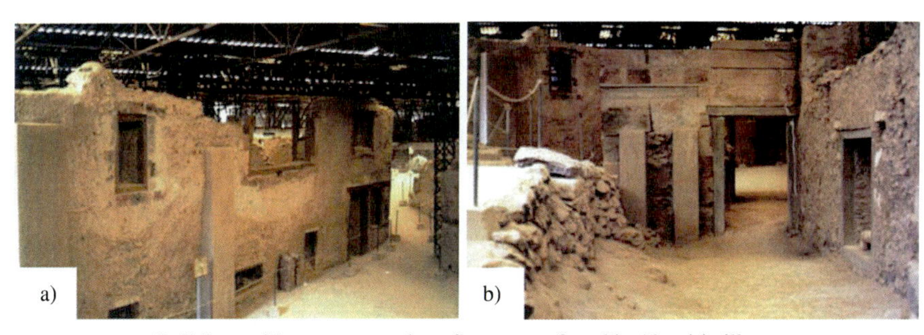

Buildings with two or even three floors were found in Akrotiri village

The choice of this particular position shows that it was a village whose inhabitants combined agriculture with fishery. The proto-Cycladic (3rd millennium BC) settlement is located under the ruins of the more recent city, which makes the assessment of its extent impossible. In places remains of proto-Cycladic walls have been located, while carved rooms inside the rock are found at several locations, probably tombs that were used by later inhabitants as stores. Thus, Santorini, having been incorporated in the Cycladic civilisation, followed the same course as the other islands during the proto-Cycladic period, and the Neolithic village evolved into a significant centre.

Intact household was found in Akrotiri settlement

Before the Minoan volcanic explosion, earthquakes destroyed part of the Akrotiri settlement and warned people to abandon the settlement

Archaeological data proves that during the last centuries of the 3rd millennium, Akrotiri had already adopted a urban character and its port was increasingly important for communication with the rest of the Aegean Sea. There is a possibility that this rapid evolution is linked to the abandonment, for unknown cause, of Poliochni, a city in Limnos referred to 1,000 years earlier. With Poliochni missing, no communication was possible between the Aegean and Pontus Euxinus, from where metals were supplied, especially from Colchis. Akrotiri soon took over Poliochni's role – due to its favourable position between Crete and the Greek mainland. At this time, goods were not being transferred between the Aegean and the Pontus Euxinus; instead trade was between the Aegean and Cyprus. This would explain how a settlement, Akrotiri, located in a small arid island, became a rich city, but it is just a possibility, since no archaeological data confirms it yet.

During the mid-Cycladic period (2,000–1,600 BC) Akrotiri prospered. This is the era of all the residencies dug up by Marinatos and of the well known frescoes; seen nowadays in the National Archaeological Museum.

Frescoes found in Akrotiri settlement and they are now kept in the National Archaeological Museum of Greece

Akrotiri rose to be one of the most cosmopolitan merchant ports of the Eastern Mediterranean, but it was destroyed during the late-Cycladic phase (1,600–1,500 BC) and then abandoned.

Palaeogeography

There is a theory suggesting that the Minoan explosion alone caused the creation of the current caldera, (Marinatos, 1972; Pichler and Friedrich, 1980). According to this theory, Minoan Santorini was an almost round island, named Strogili with an average altitude of 500–600 m. However, calderas form due to collapsing, so the volume of the collapsing material should be equal to the volume of the caldera. This fact raised the first suspicions concerning the phases responsible for the formation of the current caldera. In order to prove the above-mentioned theory, the quantity of the collapsing magma should be equal to the volume difference between Strogili and modern Thera. Measurements made on the island as well as drilling specimens of Minoan tephra from the Eastern Mediterranean, showed that the tephra ejected during the Minoan explosion was somewhere between 8 and 13 km^3 (Watkins et al., 1978). Other researchers calculated the quantity of Minoan tuffs at 31 km^3 (Pyle, 1990). In further research the volume of the caldera was estimated to be 60 km^3

(Pichler and Friedrich, 1980). Many different studies have taken place from time to time, usually leading to contradictory results. The possibility of a northern caldera in the Santorini area, flooded with sea water during Minoan times should be noted. This hypothesis is mainly based on observations of the interior inclinations on the edge of the caldera in Therassia and Oia (Heiken and McCoy, 1984).

Views of the current caldera in Santorini island complex

Topographically, Minoan Santorini was, up to a point, the same as modern Santorini. For instance, the mountains of Profitis Ilias, Mesa Vouno and Mikros Profitis Ilias were, and are still, characteristic of the Minoan relief. In some parts the Minoan relief has been dramatically altered by the deposition of tuffs of significant thickness in the valley beds, as well as on the neighbouring coastline; in some areas this spreads for more than a kilometre, e.g. in Monolithos. However at a smaller scale, topography varies from place to place. For instance, the northern half of the island is dominated by volcanic cones that have strongly influenced the thickness of the tuff deposits and given it a rougher appearance than in other parts of the island. The edges of the caldera show an incredible resemblance to the rocky site that is drawn on the Anoixi (Spring) fresco, found in the Minoan city of Akrotiri. Of course, small scale features have completely disappeared under the layer of pumice. Small valleys like the ones located under Oia and low hills like the ones uncovered in the Megalochori quarry (where the art of pottery-making had been developed in what was probably a colonised city) have vanished, leaving a new ground surface relatively flat and without any special characteristics. The only elements interrupting this flat surface come from deep erosion valleys and human land use (Gournellos et al. 1995).

Additional evidence of Thera's palaeotopography may derive from the relationship between the middle tuff series and the Minoan sequence. On southern Thera, the middle tuff series comes, for the most part, from a phreatomagmatic eruption and forms hilly depositions located on cliffy slopes in the inside of the caldera. These elements combined with the middle tuff series distribution, reveal a 6 km diameter area of depression, now inundated by the sea. This caldera was most probably formed during the explosion that produced the lower pumice series, a tuff sequence fairly similar in size and origin to the Minoan tuff.

There are many theories concerning the old coastline in the Akrotiri area. One theory suggests that the prehistoric coast was located about 800 m towards the south (Marinatos, 1972). A rather different theory suggests that the shore was located 50 m inland from the current shore and that the sea was also covering the (modern) Agios Nikolaos plain. That is where the city's port is believed to have been located (Doumas, 1983).

Part of the caldera in Oia area (northern part of Thera island)

There is evidence demonstrating the divergence between the modern and the past coastline in certain areas. Such examples can be found around the high masses of

Mesa Vouno and Mikros Profitis Ilias. Everywhere else, the Minoan coastline has disappeared under very thick tuff depositions and the current coastline lies towards the sea. The previous cliff-like slopes still remain and can be seen on the road from Perissa to Kamari and around the crystalline limestone mass in Gavrilos. It is certain that on the pre-Minoan island the modern embayment seen from Perivolos was at least partly inundated by the sea, and the area of the modern city of Emporio was closer to the Minoan coastline. This fact could support the theory that the city of Emporio was probably an important trade centre during the Minoan Age (Aston and Hardy, 1990). Despite the fact that the Minoan coastline has been covered by tuffs, there are currently two places where Minoan slopes appear. Those are in the southwestern part of Therassia and under the city of Oia on the northwestern part of Thera. The slopes facing the Akrotiri peninsula must have vanished due to the enlargement of the caldera during the Minoan explosion. The visible series are evidence of slopes less steep than the ones currently existing in some areas.

Nowadays, there are many wide and relatively flat areas, such as the area near the airport and the coastal land strip on the north. During the Minoan Age the area north of Oia and the western side of Therassia clearly had little to no chance of being available for colonisation. It is possible that the Minoan coastline was 1–2 km beyond the present coastline and the land surface followed the slope that appears on the 100 m contour.

The area west of Akrotiri must be the place where the prehistoric port was located. Around the Profitis Ilias massif there were probably other ports, as at Kamari and near Emporio. It is considered probable that the sheltered caldera served as a port (Doumas, 1983).

As far as land use in the Minoan Age is concerned, it has been discussed by many researchers (Wagstaff, 1978, Hope Simpson and Dickinson 1979; Doumas 1983). The investigation of those areas has inevitably returned very poor results due to the fact that the probably pre-existing area does not exceed 72% of that currently seen. Settlements have been recognized by buildings, walls, graves and pottery items found beneath pumice layers as well as by pieces of pottery found on exposed palaeo-surfaces. Statistically it is rather unexpected that settlements have been discovered on the island, given the fact that only a small piece of the original island still stands and only a very small percentage of the palaeo-surface is exposed. There is a high possibility that significant number of yet unknown settlements exist on the rest of the island. The density of the known settlements, as in the Akrotiri area, could possibly occur elsewhere. In geologically similar islands like Milos (Renfrew and Wagstaff, 1982) the existence of almost 20 settlements, during the Bronze Age, on an area of 33 km^2, has been proven. Proportionately to the island of Milos, in Minoan Thera 62 settlements are expected, 39 of which would probably have been preserved.

It's undoubted that systematic and intensive fieldwork will bring to light new settlements, even in the relatively limited areas available. For instance, the existence of settlements in relatively flat areas of the Minoan island, such as the slopes north of Oia and on the western part of Therassia, is considered highly probable.

The Minoan relief has been altered dramatically by the thick tuffs deposition all around the island

References

Aston, M.A. & Hardy, P.G. 1990. The Pre-Minoan Landscape of Thera: A Preliminary Statement, *Thera and the Aegean World II*, Vol. 2, Earth Sciences, The Thera Foundation, London: 348–361.

Doumas, C. 1983. *Thera: Pompeii of the Ancient Aegean*. Thames and Hudson, London.

Fytikas, M., Kolios, N. & Vougioukalakis, G. 1990. Post-Minoan volcanic activity on the Santorini volcano. Volcanic hazards and risk, forecasting possibilities. In Hardy D.A., Keller J., Galanopoulos V.P., Flemming N.C. and Druitt T.H. (Eds.) Thera and the Aegean World III, *Earth Sciences* Vol. 2, The Thera Foundation, London: 183–198.

Galanopoulos, A.G. 1958. Zur Bestimmung des Alters de Santorini – Kaldera. *Annales géologiques. Pays Helléniques*, 9: 185–188.

Galanopoulos, A.G. 1971. *The Eastern Mediterranean Trilogy in the Bronze Age, reprinted from the Acta on the 1st international Scientific Congress Volcano Thera 1969*: 184–210.

Gournellos, Th., Vaiopoulos, D., Vassilopoulos, A., & Evelpidou, N. 1995. Results of quantitative analysis in the geomorphology of Santorini Island, Proceedings of the 4th Hellenic Geographical Conference, pp. 128–139.

Heiken, G. & McCoy, F. 1984. Caldera Development During the Minoan Eruption, Thera, Cyclades, Greece, *Journal of Geophysical Research*, 89: 8441–8462.

Hope Simpson, R. & Dickinson, O.T.P.K. 1979. *A Gazetteer of Aegean Civilisation in the Bronze Age: The Mainland and Islands*. Paul Åströms Förlag, Göteborg.

Lacroix, M.L. (1896). *Sur la gonnardite. Bulletin de la Societé Française de Minéralogie*, 19: 426–427.

Lagios, E., Tzanis, A. & Hipkin, R. 1989. Surveillance of Thera Volcano, Greece: Monitoring of the local gravity field, Thera and the Aegean World III – Vol. 2, 3rd International Congress 1989.

Marinatos, S. 1939. The Volcanic Destruction of Minoan Crete, *Antiquity*, 13 (52): 426.

Marinatos, S. 1968. *Excavations at Thera. First Preliminary Report (1967 Season)*. Athens.

Marinatos, S. 1972. Thera: Key to the Riddle of Minos. *National Geographic*, 141 (5): 702

Ninkovich, D. & Heezen, B.C. 1965. Santorini Tephra. In Whittaard W.F. & Bradshaw R. (Eds.) *Submarine Geology and Geophysics*. Butterworth, London, 413–453.

Padang, M.N. 1936. Die Geschichte des Vulkanismus Santorins von Ihren Anfanfen bis zum zerstorenden Bimssteinausbruch um die Mitte des 2, Jahrtausends v, Chr, In H. Reck (Ed.), *Santorin, Der Werdegang eines Inselvulkans und sein Ausbruch 1925–1928* 1: 1–72

Pyle, D.M. 1990. New estimates for the volume of the Minoan eruption. In: Hardy D (Ed.) *Thera and the Aegean world III*, Vol 2, Thera Foundation, London: 113–121.

Pichler, H. & Friedrich, W.L. (Eds.), 1980. Mechanisms of the Minoan eruption of Santorini, In C. Doumas (Ed.), Papers and Proceedings of the Second International Scientific Congress on Thera and the Aegean World III, Vol. 2, Earth Sciences, Proceedings of the Third International Congress, Santorini, Greece: 3–9 September 1990, D.A. Hardy with J. Keller, V.P. Galanopoulos, N.C. Flemming, T.H. (Eds.), Druitt, The Thera Foundation, London.

Renfrew, A.C. & Wagstaff, M. (Eds.) 1982. *An Island Polity: The Archaeology of Exploitation in Melos*, Cambridge: Cambridge University Press.

Skarpelis, N. & Liati, A. 1987. Granite intrusion, skarn formation and mineralization in the metamorphic basement of Thera, Cyclades, Greece. 5th Meeting European Geological Societies, Dubrovnic.

Vassilopoulos, A., Vaiopoulos, D. & Evelpidou, N. 2002. Developement of a G.I.S. for the program called Edunet: Pan-Hellenic network for education, Proceedings of the congress 'Social use and spatial data: European and Greek experience in G.I.S., CD-Rom.

Velitzelos, E. 1990. New palaeobotanical data for the evolutionary history of plants in the Aegean area, with special reference to the palaeoflora of Thera. In Hardy D.A., Keller J., Galanopoulos V.P., Flemming N.C. and Druitt T.H. (Eds.) Thera and the Aegean World III, *Earth Sciences* Vol. 2, The Thera Foundation, London.

Wagstaff, J.M. 1978. A possible interpretation of settlement pattern evolution in terms of a catastrophe theory, Transactions, Institute of British Geographers, NS 3: 165–178.

Watkins, N.D., Sparks R.S.J., Sigurdsson H, Huang, T.C., Federman, A., Carey, S. & Ninkovich, D. 1978. Volume and extent of the Minoan Tephra from Santorini volcano: New evidence, from deep sea sediment cores. *Nature*, 271: 122–126.

The Petrified Forest of Lesvos A Unique Natural Monument

Nickolas C. Zouros

The Petrified Forest of Lesvos covers an area of 15,000 ha and has been declared a Protected Natural Monument. Fossil sites with standing and lying petrified tree trunks are found in many localities on the western part of Lesvos Island. The Petrified Forest was developed during Late Oligocene to Lower-Middle Miocene, due to intense volcanic activity in the area. In order to protect the Petrified Forest and ensure its proper management, serious efforts have been made during the last decades, including the foundation of the Natural History Museum of Lesvos Petrified Forest, scientific research, geoconservation, site protection measures etc. All these elements comprise the main parameters for the operation of the Western Lesvos Geopark, a body whose aims are the protection of the geological heritage and sustainable local development. The Lesvos Petrified Forest geopark comprises the famous fossil sites of the Lesvos Petrified Forest as well as a variety of other important volcanic geosites. It also includes the establishing of a network of walking trails linking geosites of interest, creation of relevant information points and eco-tourism infrastructure as well as the organization of exhibitions, scientific events and congresses and environmental education programmes and activities.

Introduction

Located in NE Aegean Sea, Lesvos Island is one of the largest Greek islands, with an area of 1630 km^2. On the western coast of Lesvos, where the volcanic rocks meet the azure blue of the Aegean Sea, natural erosion has slowly revealed the petrified remains of plant life of the distant past. No description can do justice to the brilliance, the beauty and the vivacity of their colours, the real glory of the standing fossilized trunks or the wild beauty of the volcanic landscape.

The most noteworthy concentrations of petrified trunks, making up the renowned "Petrified Forest", are located in the western peninsula of Lesvos between Sigri,

N.C. Zouros (✉)
Department of Geography, University of the Aegean, GR–81100; Natural History Museum of the Lesvos Petrified Forest, Lesvos, GR–81112, Greece
e-mail: nzour@aegean.gr

N. Evelpidou et al. (eds.), *Natural Heritage from East to West*,

DOI 10.1007/978-3-642-01577-9_2,

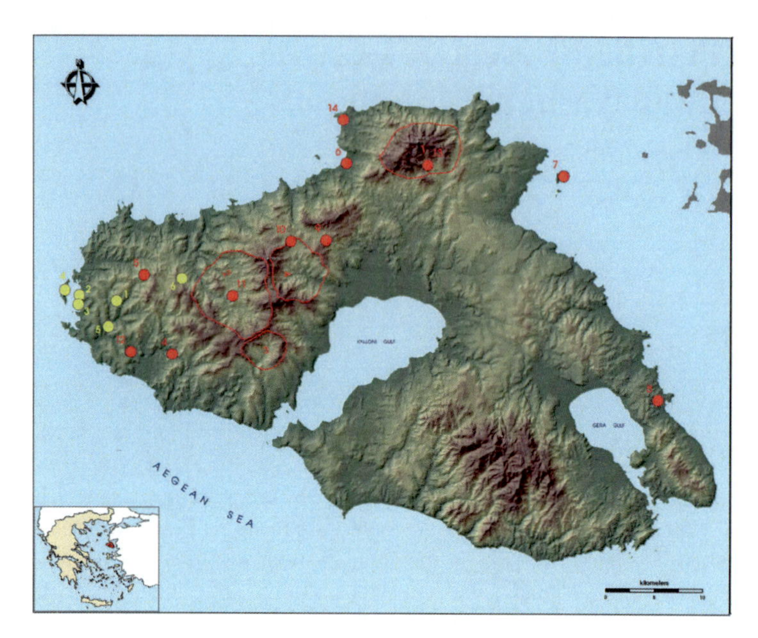

Morphological map of Lesvos island with the main fossil sites of the Lesvos Petrified Forest (yellow dots), the volcanic calderas (red line) and the interpreted volcanic geosites (red dots). Fossil sites: 1. Petrified Forest park, 2. Sigri park, 3. Plaka park, 4. Nisiopi isl. park, 5. Chamandroula park, 6. Skamiouda park. Volcanic Calderas: 1. Lepetymnos, 2. Vatousa, 3. Agra, 4. Anemotia. Volcanic geosites: 4. Eresos dome, 5. Ipsilou dome, 6. Petra volcanic neck, 7. Panagia isl. columnar lavas, 8. Alyfanta dyke, 9. Filia dyke, 10. Anemotia columnar lavas, 11. Hidira dome, 12. Eresos lakolith, 13. Pelopi columnar lavas, 14. Molyvos dome

Antissa and Eresos villages over an area of 15,000 ha. As well as petrified trunks, one encounters perfectly preserved petrified roots, fruit, leaves and seeds.

The large number of standing petrified trunks with their root systems intact and in full development provides proof that these trees were petrified in their original growing position. In other words this is an autochthonous petrified forest.

Recognising the major environmental, geological and palaeontological value of the site, the Greek State has declared the Petrified Forest to be a preserved Natural Monument (Presidential Decree 443/85).

In order to deal with the study, research, preservation, conservation and protection of the Petrified Forest, the Museum of Natural History of the Petrified Forest of Lesvos was founded in 1994. The Museum is located in Sigri villages and coordinates all the research, educational and geotouristic activities in the Petrified Forest protected area.

Due to its great geological and ecological value, a great part of western Lesvos (16,600 ha) is included in the list of the "Natura 2000" areas of Greece under the name "Petrified Forest – Western Peninsula of Lesvos". Furthermore, due to the significant presence of rare types of birds, the area is also included in the list of the most important bird habitats of Greece.

The year 2000 marked the establishment of the European Geopark Network, its objective being the cooperation of geologic parks and monuments at a European level for the development of geotourism. The Lesvos Petrified Forest is a founding member of this network. In 2001, the Museum was awarded the Eurosite Management Award for its effective management of the Lesvos Petrified Forest. In February of 2004, the Petrified Forest of Lesvos joined, the Global Geopark Network of UNESCO.

Volcanic Activity and the Creation of the Petrified Forest

The creation of the Petrified Forest is related to the intense volcanic activity that took place in the Northern Aegean during Lower Miocene. Neogene volcanic rocks dominate the central and western part of the island. Lesvos is part of a belt of late Oligocene to middle Miocene calc-alcaline to shoshonitic volcanism of the northern and central Aegean Sea and western Anatolia. The main volcanic sequence consists of andesite, dacite, and basalt lavas, ignimbrites, and a thick pyroclastic sequence (Pe-Piper and Piper, 2002).

In the central part of the island a series of volcanic centres is located along a SW-NE axis. There are impressive volcanic domes, large dikes, volcanic necks and numerous other volcanic sites. Major craters are located in central Lesvos in the regions of Vatoussa, Agra and Lepetimnos.

Early Miocene volcanic eruptions resulted in the flow of pyroclastic material that covered the vegetation.

Gigantic petrified trunk, an ancestor of today's Sequoia. This is the largest known standing trunk of a petrified tree in the world. The trunk stands 7.02 meters high and has a circumference of 8.58 meters. Here we can see the very well-preserved lower part of the trunk. The visible root system at the base of the trunk is proof that the tree is still in the same spot that it was 20,000,000 years ago. Prior to petrification the tree would have been over 100 meters high. This is trunk belongs to the species *Taxodioxylon albertense*, which is the ancestor of the *Sequoia sempervirens* now found on the west coast of the United States (California and Oregon) along the Pacific where the necessary climatological conditions (humidity) prevail for this species to grow. This species along with the related *Taxodioxylon gysaceum, Sequoia abietina, Taxodioxylon pseudoalbertense, Cunninghamia miocenica* species formed the Sequoia forests of the Aegean during the Tertiary period

In the course of these eruptions, very large amounts of lava, ash and other material were spewed into the atmosphere thereby covering extensive areas. All plant life was engulfed by a thick layer of volcanic ash. The volcanic ash and heavy rainfall following the volcanic eruptions resulted in huge mudflows of pyroclastic material, which moved from east to west covering a dense rich forest which existed on the western part of the island.

With considerable speed, these pyroclastic materials covered the trunks, branches, fruits and foliage of the forest trees.

The isolation of the plant tissue from external conditions and intense hydrothermal circulation of fluids rich in silica allowed for the perfect petrification of plant tissue under optimal conditions.

An ancestor of today's Sequoia (species *Taxiodioxylon gypsaceum*). The most characteristic of standing petrified trunks in the Petrified Forest Park, this trunk was the symbol of the park for numerous years. With its very well-preserved bark and internal structure, the trunk was revealed as a result of the natural erosion of the volcanic rocks. The external surface of the trunk is so well-preserved that it gives the impression of an aged tree rather than a fossil. Its height is 4.50 meters and circumference measures 3.70 meters

In the course of this process, organic plant material was replaced, molecule by molecule, with the inorganic material of the hydrothermal fluids. Thus, the morphological and anatomical characteristics of the trees such as the external surface, the annual rings and wood internal structures were excellently preserved.

Today, the natural erosion of volcanic rock has revealed impressive standing and fallen tree trunks measuring up to 20 m in length. Trunk diameters measure up to 3 m. Well-preserved root knots indicate that the forest is autochthonous (in its natural location), a fact which defines the uniqueness of the Lesvos Petrified Forest.

Plant Species of the Petrified Forest

Systematic study of the petrified trunks, leaves, fruits and seeds has led to the identification of the genera and species of the plants which made up the forests of Lesvos 20,000,000 years ago (Velitzelos and Zouros, 2006). The fossilized flora consists of pteridophytes, conifers, angiosperms, (mono- and dicotyledons).

Lying petrified trunk in the Plaka Park at Sigri. It was revealed by the natural weathering of the volcanic rocks which surrounded it. It is 14 meters long and the greater part of it extends into the sea. The external surface of the trunk retains all the characteristic features of the tree and the endings of its branches

The conifers are represented by the families of *Protopinaceae, Pinaceae, Cupressaceae Taxodiaceae, Gingoales*. A large number of fossilized trees in western Lesvos belong to the *Taxodiaceae* family. They are the ancestors of the present day Sequoia, considered the largest plant organism on earth, reaching heights of over 100 m and currently growing on the west coast of the United States. The *Protopinaceae* family is also represented by a large number of tree trunks located at the Lesvos Petrified Forest Park.

Of the monocotyledons, various types of palms have been identified while of the broadleaves (angiosperm-dicotyledona), species of poplar, laurel, cinnamon, plane, oak, lime, beech, alder and maple have been found.

Based on the composition of the palaeoflora, we conclude that 20,000,000 years ago in the Aegean region there were mixed conifer forests with broadleaves and palms at lower elevations. This is helpful in understanding the palaeomorphology of the forest at that time. The composition of the petrified flora indicates that the Petrified Forest developed in a subtropical climate. This changed suddenly into a continental climate, with plants characteristic of the subtropics of Southeast Asia and America.

In order to study the petrified trunks, researchers utilize the same methods used for studying the anatomy of present day trees. The macroscopic characteristics of the wood are studied, such as growth rings, pith, xylem rays, resin ducts etc.

Next, incisions are made in a piece of the fossilized trunk, and thin slices in the radial, tangential and traverse plane are prepared for microscopic study of the interior wood structure such as the cell structure, fibers and other microscopic characteristics which differ between species of trees.

The Lesvos Petrified Forest Geopark

The Lesvos Petrified Forest Geopark comprises a core zone (15,000 ha of the Petrified Forest protected area) and a broad buffer zone (more than 20,000 ha of the central volcanic terrains).

A strategic plan for the sustainable development of the area has been executed in order to link the protection and promotion of geosites with the development of geotourism. This plan takes into consideration the results of the research and excavations in the petrified forest area, the presence of important geosites (i.e. volcanic structures, domes, craters, and thermal springs) and biological reserves, the existence of spot interventions and infrastructures as well as local economic activities. Geosites are the essential elements of the Lesvos Geopark, providing information to Geopark visitors on the importance of the geological and geomorphological processes for the development of the area. Thus local people also realise that certain "rocks" in the vicinity of their houses represent remnants of outstanding phenomena and processes and demonstrate the geological history of their environment. In this way certain rock formations gain a new significance for the people and become objects to be respected and protected.

The Lesvos Petrified Forest Geopark integrates the range of resources found in the broader region, including the existing geological tourist attractions (the petrified forest park and the museum), the new parks created at important fossil sites (the Sigri park and the Plaka park), the various interpreted geosites, enchanting landscapes, wetlands, sites of natural beauty and ecological value, as well as cultural monuments (the Sigri castle, the Ypsilou monastery, the Eressos acropolis), picturesque villages, traditional gastronomy and local products.

A broad range of activities make up the main components for the operation of Lesvos Geopark, including scientific research, the creation of the geosite inventory and map, the protection, interpretation and promotion of geosites, the conservation of fossils, the creation of visiting parks, the establishment of a network of walking trails linking geosites to ecotourism infrastructures, the development of environmental education programmes on geosites, the organisation of scientific and cultural events, and the promotion of monumental geosites.

The Lesvos Petrified Forest Geopark applies certain management measures for the protection, conservation and promotion of the inventory of geosites present in

Pine (*Pinoxylon paradoxum*). One of the most characteristic petrified trunks in the Petrified Forest Park discovered in its original position. This is a conifer of the Protopinaceae family and it is characterized by extremely well-preserved bark, internal structure and root system. The trunk is 1.50 meters high and has a circumference of 4.20 meters. In the trunk's interior structure are impressive large annual rings with a number of concentric groups, a characteristic feature of this species. At the base of the trunk is a root system in full development attesting to the fact that the Petrified Forest of Lesvos is autochthonous, meaning that the trees were located in the same position prior to petrification. The autochthonous nature of the forest renders this petrified monument truly unique. Recent excavations revealed a 6-meter trunk lying adjacent to the root system of the pine trunk. The external surface of the lying trunk is particularly well-preserved as are the morphological characteristics and structure of the wood. The bark also displays branch and twig knots

the territory. These measures comprise: (1) regular maintenance (fencing, cleaning) and janitor services to protect geosites against abuse and vandalism; (2) geosite monitoring providing the necessary measures and protective installations against weathering and erosion; (3) treatment of vulnerable geosites with annual conservation and protective measures (preparation, sealing); (4) interpretation of geosites with onsite information panels, leaflets and field guides, and; (5) organisation of on-site activities (education programmes, guided trekking on geosite trails, various recreation activities in the vicinity of geosites, etc.) that assist in raising public awareness about the importance of geosites.

The main infrastructure for geotourism in the Lesvos geopark is the creation of "lava paths" that invite the visitor to follow the ancient path of the pyroclastic flows from the main volcanoes to the petrified forest. These are footpaths that link the various geosites and other sites of interest throughout the geopark. Panels along the footpaths provide information about the different geosites that the visitor will encounter along the way. On entering the Lesvos geopark region, signs along the Mytilene-Kalloni-Sigri road direct the visitor towards the Petrified Forest and demarcate the borders of the protected area. Walking trails start from different points along the main road. The Lesvos geopark has also created links with local tourist enterprises, handicrafts and women cooperatives producing local food and drinks.

A Visit to the Petrified Forest

Visitors to Lesvos may be offered a number of alternatives if they wish to visit the Petrified Forest. Firstly, they may go to Sigri, to visit the Natural History Museum, where the displays include finds from the Petrified Forest and an account of the geological history of the evolution of the Aegean, from the processes which led to the creation of the Petrified Forest 20 million years ago up to the present-day ecosystems. Then they can visit the open-air parks; The Petrified Forest park , the Sigri park, the Plaka park and the small island of Nisiopi.

There have also been paths created for those who love walking, which connect the sites of fossils. Informative signs have been placed in the most important spots along the paths.

The Sigri park covering an area of 2 ha it has been established immediately adjacent to the Museum and contains unique concentrations of roots of fossilised conifers, still in their original position.

From Sigri visitors can visit by boat the picturesque Megalonisi or Nisiopi islet, where they will be able to enjoy the spectacle of very fine land and marine sections of the Petrified Forest. Fossilised trunks of angiosperms and conifers with a great variety of colours lie displayed on the islet's west coast.

The Plaka park is located 800 m south of Sigri, covering an area of 7 ha. There the visitor can admire impressive petrified tree trunks, unique fossilized leaves and of course the thickest standing tree trunk in the world with a perimeter of 13.7 m.

Tetraclinoxylon velitzelosii
An ancestor of today's Cypress. This petrified trunk belongs to the Cypress tree family and is identified as a new species *Tetraclinoxylon velitzelosii*. The upper part of the trunk has broken off and appears on the ground immediately adjacent to the standing section. Wood characteristics such as annual rings are clearly visible. The trunk is 4.55 meters long with a diameter of 1.20 m

A visit to the Petrified Forest Park, with its of total area 28.6 ha, is a unique experience. You can access it through turnpike road 5 km from the 9th km of the provincial road from Antissa to Sigri. Visitors are astounded by the host of perfectly preserved fossilised trees, making up a whole forest ecosystem from the distant past. Nowhere else on the planet are there so many standing fossilised trunks.

Visitors will be struck by the exceedingly complex colour combinations with which nature has adorned the petrified trees, and will marvel at the "authenticity" of the fossilised trunks, which have retained the tiniest detail of their interiors. Visitors will be impressed by the number and dimensions of the dozens of standing and fallen trunks; these are scattered over the area and are sometimes more than 7 m in height and 22 m long. The biggest is 7.02 m in height and is the tallest in Europe.

The Museum

The museum is located on the top of a small hill that offers an excellent view of the surrounding area, the picturesque town of Sigri, and the sea. Its one-hectare grounds border the Sigri Park to the south. The single storey stone structure of the museum is built of the gray lava that abounds in the region.

Museum of Natural History of the Petrified Forest of Lesvos

Once inside the museum, visitors can view the fascinating permanent exhibits. The first exhibition room, "The Petrified Forest Hall", presents the evolution of plant life on earth from the appearance of the planet's first single cell organisms to developed plant life and the creation of the Petrified Forest. The flora of the Petrified Forest is presented with fossil remains of over 40 different species found and identified in the broader area of western Lesvos. Petrified trunks, branches, twigs, impressive petrified leaves, leaf imprints, fruit and roots are displayed in front of large-sized pictorial depictions of the plants they represent. Characteristic examples from other fossil-bearing sites in Greece (Kymi and Aliveri on Evia Island, Vegora, Elassona, Santorini Island) are also part of the exhibition. This hall also houses the first evidence of the existence of animals living in the Petrified Forest, such as the fossil jawbone of a dinothere (*Predinotherium bavaricum*), a trunked ancestor of the elephant from the region of Gavatha, Antissa, dating back 20 million years. This find constitutes one of the oldest fossils of a vertebrate in Greece and is particularly rare in Europe (Koufos et al., 2003).

In the second exhibition room, entitled "The Evolution of the Aegean Hall", there is a presentation of the various geological phenomena and processes associated with the creation of the Petrified Forest and the general geological history of the Aegean basin over the last 20 million years. Among the topics presented are the movement of tectonic plates in the region, the subduction of the African tectonic plate and the evolution of volcanic activity in the area. The exhibit incorporates impressive examples of volcanic rock and formations. Models of volcanoes and the stratigraphy of volcanic products on western Lesvos are also part of the presentation. Embedded in the volcanic rock we find perfectly preserved petrified trunks, roots, fruits, leaves and seeds. Furthermore, a paleogeographic reconstruction depicting the development of the region from the Tethyan Ocean to the continental Aegis mainland and the creation of the Greek Archipelago is included in the exhibit. The active stress pattern of the Aegean region, the active volcanoes, the seismic faults and the numerous geological monuments depicted by the exhibits, remind us that the geological processes which led to the creation of the Petrified Forest have not yet come to an end.

The Museum also offers an audiovisual presentation "Petrified Forest: The Pompeii of the Plants" which is a tour through the geodynamic phenomena that led to the creation of the Petrified Forest and the wealth of discoveries that constitute this unique monument. A second audiovisual presentation entitled "Biodiversity of the Northern Aegean" presents a journey through today's ecosystems and the wonderful natural wealth of the northern Aegean islands. The latter was created with the scientific support of the Department of Environmental Sciences of the University of the Aegean.

In addition to the permanent exhibitions, the Museum organizes exhibitions and public presentations on the ecological value of the Petrified Forest protected area, and fauna and flora.

Many special temporary exhibitions in various fields of Paleontology, Geology and Physical History as well as art exhibits (paintings, photo, sculpture etc) take place every year

The Museum also organises special environmental education programmes in order to cultivate amongst young students a widespread sense of respect for the Earth's heritage, natural monuments and the environment.

References

Pe-Piper, G. & Piper, D.J.W. 2002. The igneous rocks of Greece. The anatomy of an orogen, Gebruder Borntraeger, Berlin, Stuttgart: 573.

Koufos, G., Zouros, N., Mourouzidou, O. 2003. *Prodeinotherium bavaricum* (Proboscidea, Mammalia) from Lesvos island, Greece; the appearance of deinotheres in the Eastern Mediterranean *GEOBIOS 36,305–315.*

Velitzelos, E. & Zouros, N. 1998. New results on the petrified forest of Lesvos. *Bulletin of the Geological Society of Greece*, 32/2: 133–142.

Velitzelos, E. & Zouros, N. 2006. The petrified forest of Lesvos. Topio publications, Athens: 144.

Velitzelos, E., Mountrakis, D., Zouros, N. & Soulakellis, N. 2003. Atlas of the Geological Monuments of the Aegean. *Ministry of the Aegean*. Athens: Adam Editions: 352. (in Greek)

Zouros, N. 2004. The European Geoparks Network. Geological heritage protection and local development. *Episodes*, 27/3: 165–171.

Zouros, N., Soulakellis N., Mountrakis, D. & Velitzelos, E. 2004. Mapping, classification and assessment of geotopes in the Aegean. Proceedings of the 7th Hellenic Geographical Conference, Geographical Society of Greece, 1: 527–534.

Zouros, N. 2005. Assessment, protection and promotion of geomorphological and geological sites in the Aegean area, Greece. Géomorphologie: relief, processus, environnement, no 3: 227–234p.

Zouros, N. 2007. Geomorphosite assessment and management in protected areas of Greece. The case of the Lesvos Island – coastal geomorphosites. *Geographica Helvetica Jg* 62, Heft 3: 169–180p.

The Lavrion Mines

Athanassios Katerinopoulos

Introduction

Lavrion Municipality extends 15 km along the SE coast of Attica, about 50 km from the center of Athens (http://en.wikipedia.org/wiki/Laurium). Next to Lavrion there is cape Sounion and the well-known temple of Poseidon. Ore exploitation of the Lavrion area has a history of 5,000 years. The naval strength of Athens in the fifth century BC and the silver decorations of its buildings are indications of the riches of the Lavrion mines. Apart from the ancient silver mining, Lavrion is known worldwide for the variety and the beauty of its mineral samples. These can be still found underground in the galleries, but also on the surface, wherever mining waste has been left.

Kyanotrichite, Lavrion

A water tank in Soureza valley

A. Katerinopoulos (✉)
National and Kapodistrian University of Athens, Department of Geology and Georнvironment, Section of Mineralogy and Petrology, Panepistimiopolis, Gr-15784, Athens, Greece
e-mail: akatern@geol.uoa.gr

N. Evelpidou et al. (eds.), *Natural Heritage from East to West*,
DOI 10.1007/978-3-642-01577-9_3,

The Ancient Mines

The great mining boom of Lavrion began in 483 BC, when the rich deposits of silver were discovered in Maronia (today's Kamariza or Agios Konstantinos). The silver that at that period "flowed like water from the spring" came from the silver mines, the famous arghyreia, from which the state and businessmen made astronomical profits in the fifth century. It is reported that a certain Kallias earned 1,200,000 drachmas (5.5 tons of silver) in 1 year (Marinos and Petrascheck, 1956).

The ancient underground galleries extend over a total length of hundreds of kilometres and were built on six levels. More than 1,000 shafts have been found in the region, some of which are as deep as 119 m. In the wider Lavrion area one can still see hundreds of gallery entrances and shafts, especially in the Plaka and Kanariza areas (Konofagos, 1980).

The galleries were hewn out using nothing but pickaxes, hammers and chisels. Fire was used to a lesser degree where explosives would be utilised today. That is, the rock would be heated, and then water was thrown on it suddenly, causing the rock to shatter owing to the abrupt contraction. It is calculated that the work of building a gallery proceeded at a rate of 12 m a month, while for shafts the rate was just 5 m a month.

The area of Lavrion is extremely arid. The lack of the water led to the construction of a drainage system bringing the water into huge tanks. Just before the main tank there was a smaller one for the precipitation of any impurity. One should point out the perfect impermeability of these tanks. Note that the same insulation material, used 2,500 years ago in Lavrion waterproof tanks, is also used nowadays all over the world for the insulation of radioactive waste storage pits. Water tanks and the remaining of the drainage system can be still seen in the Soureza valley.

Using Lavrion silver, Themistocles was able to build the powerful Athenian fleet, which together with that of the rest of Greece saved the ancient civilisation at the battle of Salamis. The Athenian triremes then numbered 200, out of a total of 314 in Greece. Every trireme cost two talents, which were equivalent to 54 kg of silver. Before the Peloponnesian war, 270 tons of silver were stored in the treasury of the Parthenon and as much again was in circulation as coinage: the Athenian silver coin "glafka" (=owl). The same owl appears on the front face of the Greek 1 Euro coin.

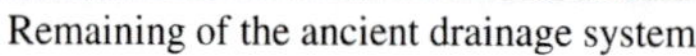

Remaining of the ancient drainage system

The glafka coin

During antiquity, the mines belonged exclusively to the Athens city-state, which granted concessions to exploit them primarily to Athenian private citizens. The concessionaire would hand over one twentieth of his profits to the state. At the period of maximum activity, this amounted to 50–100 talents a year (1 talent = 27 kg silver or 6,000 silver drachma).

State control of the operation of the mines and of their production was systematic and strict. For this reason there was special mining law and a special court to enforce it. Anyone would be punished who exceeded the limits of his concession, operated the mine that had been ceded to him in an irregular way, or jeopardised the future of other mines. It is reported that under Lycurgus (330 BC), a certain Diphilos, known for the large profits he extracted from Lavrion, was put to death for his greed, because he cut down the mineral columns that were left intact to prop up the galleries.

The Slaves

Looking at the Athenian frescos one can see figures of miserable slaves working in the Lavrion mines. These poor creatures were considered "speaking tools" and common merchandise, without any political or legal rights.

Nevertheless there were unwritten laws according which any slave, if treated cruelly, had the right to take shelter in a temple; his master was then obliged to sell him.

No master could put his slave to death without a court judgment. Also any Athenian free citizen could accuse any person of treating his slaves cruelly. In the fifth century BC the slaves in the Lavrion mines numbered about 15,000.

During the work there were guards at the entrances of the galleries and on the top of towers, to prevent the slaves from escaping. But there was a special respect for their religious customs, and even burial traditions.

Despite these conditions, there was a number of slave revolts, the most important being the one in 413 BC (during the war between Athens and Sparta) resulting in the defeat of Athenians, due to the poor state of the economy.

Lavrion then became inactive, so Pausanias (second century AD.) said "Near Sounion, at Lavrion, Athenians' silver mines existed". This continued for centuries, till 1860.

The Ore

The ancient Athenians developed a very advanced technique to enrich the ore. After the hand-picking, the material was crushed in rock mortars and brought to special "lavatories", where they separated the argentiferous galena from sphalerite, pyrite, calcite and the clay minerals. A restored lavatory can be seen at Thorikos, next to the ancient theater, one of the oldest in Greece. A characteristic of this theater is the elliptical pattern, in contrast to the semicircular pattern of later ones. There is also the oldest gallery found in Lavreotiki (3,000 BC).

Restored lavatory at Thorikos

The ancient theatre at Thorikos

Dozens of lavatories are preserved at Soureza valley, an archaeological site, about 2 km from Ag. Konstandinos, well worth a visit.

On the way to Soureza, next to the park, there is a spectacular depressed area, named "The chaos", formed by the collapse of a huge cave (http://www.mylavrio.gr).

One of the ancient lavatories in Soureza valley

The chaos

The enriched ore was transported to the furnaces in order to separate the Ag-Pb alloy. Successive layers of ore and charcoal were layered and put in fire, so as the heavier argentiferous lead was concentrated at the bottom while the lighter "slag" flowed from the top. The slag was discarded as a metallurgical waste. Huge piles of slag covered the Lavreotiki area. Today only a few remaining samples can be found, as the material was reworked in order to exploit the remaining silver content.

The separation of the Ag-Pb alloy into lead and silver fractions (cupellation) was done in special furnaces by an extremely difficult and precise procedure. The alloy was melted by the burning of wood with the help of hand-bellows. This procedure is still used in silver metallurgy.

Experienced workers melted the alloy at 950°C, when the lead oxidized to PbO flowing into a ceramic bowl, while the purified silver (not affected by atmospheric oxygen in this temperature) remained in the furnace. The procedure was considered completed when there were "flowers" i.e. bubbles of the atmospheric air appearing as white flowers on the silver surface. This silver was 98% pure and ready for coinage. From 1,000 kg of silver-lead ore they could get 200 kg of lead and 25 kg of silver. The lead was used at first to connect parts of columns, and to stabilise statues. Later it was used for drainage pipes.

The income from the Lavrion mines maintained the economic strength of the city of Athens until the rule of Alexander the Great. By the end of the second century BC, during the roman expansion, the mines were closed.

Modern Lavrion History

In 1860 the Greek mining engineer Kordelas visited Lavrion and wrote a memo to the Greek Ministry of Development, referring to the possibilities of reworking the ancient slag (Kordellas, 1993).

At the same time the Italian businessmen J. B. Serpieri recognized the ballast of a ship thrown at the Italian coast as pure smithsonite. Asking for more information he learned that the ship was returning from Lavrion-Greece, where it was loaded by this "useless" ballast.

Serpieri visited the Lavrion area and, bearing in mind the Kordelas report, he established, along with the French businessmen Ilarion Roux, the Italian-French company «Ilarion Roux et Cie» (1864). The company started exploiting some new deposits but mainly reworking the ancient slag.

It was the largest company in Greece at that time employing 1,200 workers, operating 18 Spanish furnaces of the "castilliano" type, a large turnery and their own locomotive.

But the company only had permission to exploit new ore deposits and not the ancient slag that was "a product of the human activity" and in 1869 legal proceeding were started by the Greek state. The negotiations ended 4 years later when A. Sygros bought the company and renamed it "Greek metallurgy of Lavrion". At the same time Serpieri established a new company: "Compagnie Francaise des Mines du Laurium"

The small settlement of Lavrion expanded to a city of 10,000 people. The two companies owned the houses and the shops. They also were taking care of the schools and the churches of the town, as well as pharmacies and the hospital treatment of the workers.

The Greek company was the first one in Greece that used electricity, telephone and other modern technologies. They also constructed the railway that connected Athens with Lavrion (1882–1885). The company lasted until 1917, when the slag deposits were exhausted (http://www.eranet.gr/lavrio).

The French company lasted until 1983, when it was sold to a British company that did not continue the mining.

The Lavrion Minerals

The broader mining region of Lavreotiki constitutes a natural museum, since in it can be found more than 15% of all the recognised minerals on earth, known world-wide for their variety and for their particular beauty (Katerinopoulos and Zissimopoulou, 1884).

The Athens University Museum of Minerals and Rock owns a very large collection of Lavrion minerals. All the samples are outstanding, but the pride of the collection is a unique stalactite made of smithsonite, about 70 cm long.

Part of the Lavrion minerals collection, Museum of Mineralogy-Petrology, Athens University

The stalactite of smithsonite. (Museum of Mineralogy-Petrology, Athens University)

The Slag Minerals

Despite the highly developed metallurgical techniques taking place in the ancient factories, the ancient Greeks could not take out all the silver from the ore and, after smelting, they discarded the slag, sometimes into the sea.

The sea water contains Na, Cl and other trace elements, which penetrated for 1,000 years into the slag. The chemical reaction with lead, silver and other trace elements existing in the slag resulted in the growth of perfect crystals of various minerals, some of which are very rare and exist only in Lavrion, such as Thorikosite (named from the locality Thorikos).

In winter the slag is carried out of the sea by waves and then lies scattered throughout the coast of Lavrion. Collectors come from all over the world to collect it.

Whenever you visit Greece do not miss out Lavrion – a place of history and beauty.

References

Marinos, G. & Petrascheck, W.A. 1956. Lavrio IGSR (ed.) Athens, McDonald, 248 p.

Katerinopoulos, A. & Zissimopoulou, E. 1884. *The minerals of the Lavrion mines. Gr. Assoc. Min. Fossil Collectors*. Athens: 304.
Konofagos, K. 1980. *Ancient Lavrion and the Greek technique of silver production*. Ekdotiki Ellados.
Kordellas, A. 1993. Lavrion (translation of "Le Laurium, Marseille, 1869"), Library of the society of studies of Lavreotiki No 6.
http://en.wikipedia.org/wiki/Laurium
http://www.eranet.gr/lavrio/html/glavrio.html
http://www.mylavrio.gr/index.php?option=com_content&taskviewid=41&Itemid=50

Tafoni and Alveole Formation. An Example from Naxos and Tinos Islands

Niki Evelpidou, Dimitra Leonidopoulou, and Andreas Vassilopoulos

Weathering formations resembling small caves, known by the name of Tafoni, are a characteristic, but not exclusive, feature of the Mediterranean area. Examples of such geomorphological formations have been recorded in Sardinia and Corsica (Klaer, 1956; Frenzel, 1965), in Tuscany (Martini, 1978), in S. Spain (Mellor et al., 1997) and in the Aegean Sea area (Greece) (Riedl, 1991; Hejl, 2005).

Within the Aegean Sea area, Tafoni formations have been studied in the islands of Thasos (Resch et al., 1989), Naxos (Sabot, 1978; Weingartner, 1982; Evelpidou, 2001; Weingartnerand and Wögerbauer, 2003), Tinos (Theodoropoulos, 1974; Livaditis and Alexouli-Livaditi, 2001; Maroukian et al., 2005; Leonidopoulou, 2007), Paros and Ios (Hejl, 2005).

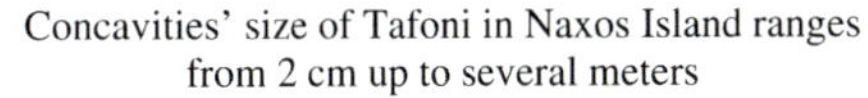

Concavities' size of Tafoni in Naxos Island ranges from 2 cm up to several meters

Tafoni form in Tinos Island

N. Evelpidou (✉)
National and Kapodistrian University of Athens, Department of Geology and Geornvironment, Section of Geography and Climatology, Panepistimiopolis, Gr-15784, Athens, Greece
e-mail: evelpidou@geol.uoa.gr

N. Evelpidou et al. (eds.), *Natural Heritage from East to West*,
DOI 10.1007/978-3-642-01577-9_4, © Springer-Verlag Berlin Heidelberg 2010

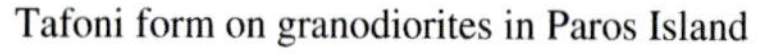

Tafoni form on granodiorites in Paros Island

Honeycomb weathering forms bigger than 0.5 m are defined as Tafoni, whereas forms smaller than 0.5 m are defined as Alveoles

The definition of Tafoni formations was established by Penck (1894), who used it to refer to the honeycomb weathering of granite in Corsica. Honeycomb weathering formations bigger than 0.5 m are defined as Tafoni, whereas formations smaller than 0.5 m are defined as Alveoles; the latter are sometimes developed (as in Tinos Island) on schist surfaces (Theodoropoulos, 1974). These formations are often characterised as "aeolian erosion formations". In fact, their development is partly due to wind action, but is mainly due to chemical weathering (Soukis et al., 1998).

Tafoni forms on greenschists in Marlas area, in northern Tinos Island

Honeycomb weathering formations on granodioritic rocks on Naxos Island, Stellida area

A wide variety of conditions may affect the formation and development of Tafoni and can be grouped in four categories (Wilhelmy, 1981):

1. mineralogical and structural properties of the bedrocks,
2. topographic elements which are independent of Tafoni formation, i.e. the exposure of the slope and pre-existent relief features,
3. the climatic characteristics of the area, and
4. the microclimatic conditions created by the Tafoni itself.

Weathering begins in weakness zones, which while broadening, may cause rock disintegration resulting to sphere shaped end products (Paros Island)

Most authors agree that Tafoni are developed especially in granites and granodiorites, as well as in gneisses and sandstones, that is in typical medium to coarse-grained silicate rocks with granular fabric (Wilhelmy, 1981; Mellor et al., 1997; Matsukura and Tanaka, 2000). A number of processes and conditions have been suggested for Tafoni weathering, including hydration, increased pressure of salts and specific microclimatic conditions (Frenzel, 1965; Martini, 1978; Kirchner, 1996; Campell, 1999; Matsukura and Tanaka, 2000; Weingartner and Wogerbauer, 2003).

Considering that Tafoni are frequent in areas that are humid, shady and protected from the wind, their creation must be connected, not only with wind action, as had originally been thought, but also mainly with water action. Weathering takes place as water infiltrates in the rock's porosity. Temperature fluctuations generate infiltrating water's contraction and dilatation, resulting in the weakening of the cohesion of the rock's surface layer.

Their absence in absolutely arid climates shows that Tafoni are not generated exclusively by intense temperature fluctuations and Aeolian weathering. On the contrary, their presence in coastal areas is due to stresses developed during the crystallisation of a chemical combination, which then dissolves. In particular, sodium chloride (NaCl) is transported, though wind and sea waves, and penetrates inside the rocks' discontinuities, where it recrystallises, resulting in an increase of stresses in the capillary cracks.

Tafoni forms on granitic rocks, in Livada Bay in SE Tinos Island

Tafoni forms on granodioritic rocks, in Kolympithres beach in NW Paros Island

According to other authors, the main cause for Tafoni creation is chemical weathering due to the kaolinisation of feldspar and elements transfer, like the removal of iron from biotite. In areas where climatic conditions are suitable saline dilutions occur on the rock's surface, due to intense evaporation. Thereby, a solid peripheral silicon, iron and magnesium oxide zone is developed, which surrounds the weathered inner part. In more humid areas, this zone breaks up or doesn't develop at all. In that case, the excavation starts from the already weathered, inner part of the rock.

Tafoni formation on granodiorite rocks, in Naxos Island

At the early stage of Tafoni development, hollows are frequently created on exposed rock or along discontinuities, along which the rock has higher porosity and reduced strength (Weingartner, 1982; Matsukura and Tanaka, 2000). Because of the Mediterranean climate, the exposed surface of the rock dries very quickly after rainfall, while the inner part of the rock and soil, found at the base of the formation, keeps its moisture longer. These conditions are more intense during spring and autumn, when rainfall is higher and the air temperatures are high enough to generate chemical weathering. However, chemical weathering is more intense at the base of the formation and is probably supported by capillary water migrating though rock's porosity, from the base and inner part of the rock to its drying exterior. On the rock surface of a developing Tafoni, a case hardening can be created by near-surface cementation, which protects the rock form further decomposition (Wilhelmy, 1981).

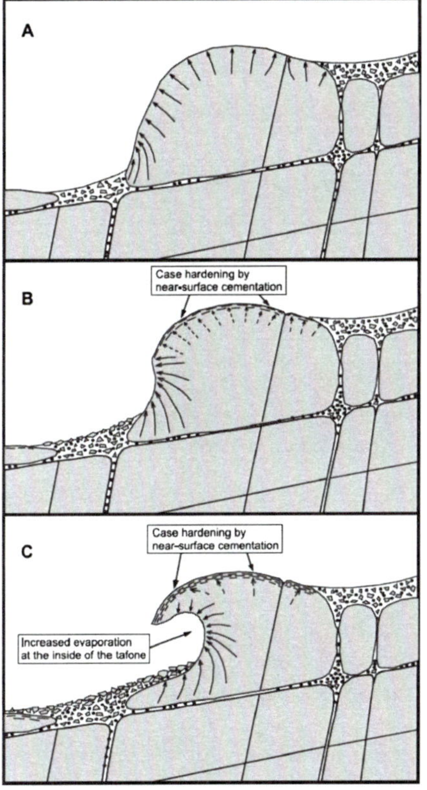

Schematic representation of tafoni development

The Lilies Fresco from room D2 of ancient Akrotiri. The arrows refer to different stages of tafoni development

Honeycomb weathering and Tafoni development has amazed humans for thousands of years, as is apparent from archaeological finds (Hejl, 2005). In particular, on the prehistoric settlement of Akrotiri, in Santorini Island, copper Age wall paintings representing Tafoni development have been discovered. In Figure 12 small sized rocks with natural vegetation and shallow concavities are shown. The shape of these rocks resembles the different evolving stages of Tafoni (Hejl, 2005).

In the Cycladic Islands Tafoni are mainly developed on crystalline rocks of medium to large grain size, but also on other rocks, such as sandstones, limestones and schists. The width and depth of these foms varies from some centimeters to several meters. Their shape tends to be ellipsoidal or spherical. Bigger Tafoni usually have thinner walls, a vaulted shape and their concavity is at the lower part, having a gentle form. Very frequently smaller rounded concavities combine with each other so as to develop a bigger concavity, especially on the roof of the Tafoni (Evelpidou, 2001).

Usually, on the walls and on the roof of a Tafoni, smaller concavities and humps are observed. These forms are developed on steep slopes and have a hemispherical roof (Theodoropoulos, 1975). The same writer suggests two types of this geomorphological form: (a) Tafoni formation on the base of the bedrock and on the weather-weakened zones of the rock, and (b) Tafoni formation having a concave side formed by bedrock fracturing. In some cases tafoni are oriented towards the lee side of the rock.

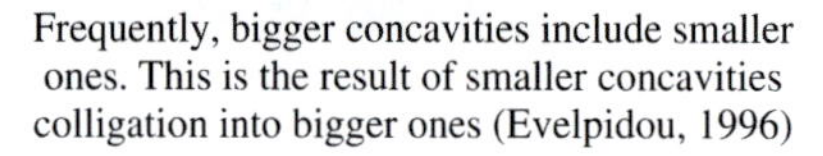

Frequently, bigger concavities include smaller ones. This is the result of smaller concavities colligation into bigger ones (Evelpidou, 1996)

Tafoni on greenshists in northern Tinos Island

In Tinos Island, Theodoropoulos (1974) traced these forms on actinolithic schists. After studying rock samples of the bedrock and weathered rock though the microscope, he concluded that there was a siqnificnt amount of calcite in the bedrock samples, while there was very little or no calcite in the weathered samples. This result, in combination with the fact that there was no degradation of the Feldspars or Chlorite leads to the conclusion that the main weathering cause must be connected with the deposition of calcite. Subsequently, mechanical weathering takes place; wind action only takes effect for the deposition of the weathered material.

Microclimatic conditions inside a well developed Tafoni are quite different than the climatic condition outside it. Measurements yield by Riedl (1995) on the granitic rocks surface in Pinakoto Mt., at 510 m altitude, in Tinos Island, are shown in Table 1:

Table 1 Temperature and air relative humidity measurements inside and outside a Tafoni (Riedl, 1995)

	Outside the Tafoni	Inside the Tafoni
Air temperature	25.5–27.4°C	24.9–23.7°C
Air relative himidity	62.5–56.4%	65.00–66.8%
Rock temperature	26.6–27.9°C	24.3–24.2°C
These measurements took place on August from 11:00 till 14:00.		

Summing up, for Tafoni oriented towards N a microclimate is developed inside the concavity, which is characterised by lower air temperature, but higher values of air relative humidity. Comparable measurements during afternoon hours showed that the air temperature inside the concavity was 1.7–4.2°C lower than it was outside, while air relative humidity inside was 2.5–10.4% higher than it was outside.

References

Campell, S.W. 1999. Chemical weathering associated with tafoni at Papago Park, Central Arizona. Earth Surface Processes and Landforms 24:271–278.

Evelpidou, N. 1996. *Geological and Geomorphological study on Paros Island, using Photo-interpretation and G.I.S.* M.Sc. Thesis, Faculty of Geology and Geoenvironment, National and Kapodistrian University of Athens. (Unpubl.). (In Greek).

Evelpidou, N. 2001. *Geomorphological and environmental study on Naxos island, using remote sensing and G.I.S.* PhD thesis, Faculty of Geology and Geoenvironment, National and Kapodistrian University of Athens. In Greek. Gaia 13. Athens: 226. (In Greek).

Frenzel, G. 1965. *Studien an mediterranen Tafoni. Neues Jahrbuch für Geologie und Palaeontologie Abhandlungen 122–3*: 313–323.

Hejl, E. 2005. *A pictorial study of Tafoni development from the 2nd millennium BC. Geomorphology 64*: 87–95.

Kirchner, G. 1996. Cavernous weathering in the Basin and Range area, southwestern USA and northwestern Mexico. *Zeitschrift für Geomorphologie N.F. Supplementband 106*: 73– 97.

Klaer, W. 1956. Verwitterungsformen im Granit auf Korsika. *Petermanns geographische Mitteilungen, Ergänzungsheft 261* (Gotha).

Leonidopoulou, D. 2007. *Geological and geomorphological factors in the formation of intrinsic vulnerability of fractured rocks.* Application on the island of Tinos. PhD Thesis. Dep. of Dynamic, Tectonic and Applied Geology, Faculty of Geology and Geoenvironment, National and Kapodistrian University of Athens. Athens: 361p. (Unpubl.) (In Greek).

Livaditis, G. & Alexouli-Livaditi A. 2001. *The morphology of the Tinos Island, Bull. Geol. Soc. of Greece, Vol. XXXIV/1*: 389–396. (In Greek).

Maroukian, H., Leonidopoulou, D., Skarpelis, N., Stournaras, G. 2005. *Effects of lithology and weathering on particle size variability of sediments in the coastal environment in se Tinos Island, Greece.* 6th International Conference on Geomorphology, Zaragoza Spain.

Martini, J. 1978. *Tafoni weathering, with examples from Tuscany, Italy. Zeitschrift für Geomorphologie N.F. 22*: 44–67.

Matsukura, Y. & Tanaka, Y. 2000. *Effect of rock hardness and moisture content on tafoni weathering in the granite of Mount Doeg-Sung, Korea. Geografiska Annaler 82 A*: 59–67.

Mellor, A., Short J. & Kirkby, S.J. 1997. *Tafoni in the El Chorro area, Andalucia, Southern Spain.* Earth Surface Processes and Landforms 22: 817–833.

Penck, A. 1984. *Morphologie der Erdoberflache*, Vol. 1: 214. Stuttgart.

Resch, Th., Stangl, D. & Weingartner, H. 1989. *Tafoniverwitterung auf Thassos*. Ein Fallbeispiel. Salzburger Geographische Arbeiten 18 (Beiträge zur Landeskunde von Griechenland III): 77–88.

Riedl, H. 1991. Beobachtungen zur Klimamorphologie von Massengesteinen in den alt- und neuweltlichen Subtropen vorwiegend des mediterranen Typs. Festschrift für Herbert Paschinger, Arbeiten aus dem Geographischen Institut der Universität Graz 30: 235–252.

Riedl, H. 1995. *Beiträge zur regionalen Geographie der Insel Tinos (Kykladen) mit besonderer Berücksichtigung des quasinatürlichen Formenschatzes. Beiträge zur Landeskunde von Griechenland V, Band 29*, Salzburger Geographische Arbeiten. Salzburg.

Sabot, V. 1978. *La geomorhologie et la geologie du Quaternaire de l' ile de Naxos, Cyclades-Greece* Thesis, Vrije University, Brussel: 128.

Soukis K., Koufosotiri E., Stournaras G. 1998. *Special landforms on Tinos Island: spheroidal weathering "TAFONI" forms*. 3rd International Scientific Symposium of Protected areas and Natural Monuments. Mytilini. (In Greek).

Theodoropoulos, D. 1975. Honeycomb weathering phenomena (TAFONI) on Tinos Island. *Annales Géol. des pays Hellén, XXVI: 149–158*. Athens. (In Greek).

Weingartner, H. 1982. Tafoniverwitterung in Naxos. Geographische Studien auf Naxos. *Salzburger Exkusionsberichte 8*: 90–106.

Weingartner, H. & Wögerbauer E. 2003. *Microclimate and Tafoni weathering. Results of a field study on the island of Naxos (Cyclades, Greece)*. Proceedings of the 6th Pan-Hellenic Geographical Conference of the Hellenic Geographical Society, Thessaloniki, Volume 1: 409–414.

Wilhelmy, H. 1981. *Klimamorphologie der Massengesteine. 2. Auflage*: 254, Wiesbaden.

Origin, Geology and Geochemistry of Mpouharia and Nohtaria Landforms, in Mikrovaltos Kozani, NW Macedonia – Greece

Akindinos Kelepertzis and Evagelos Tziritis

Introduction

The long-term and continuous change of the characteristics of the geosphere is impressed in rocks and surface formations that constitute the fundamental elements of the earth's crust. Over recent centuries several empirical observers and scientists have investigated and gathered evidence of geo-historical evolution. Extended in situ observations as well as the correct interpretation of the scattered rocks, fossils and landforms, have supplied us with adequate information, which has been crucial for the justification and establishment of models of this historical and geological evolution. In this context, the term geosite, which includes landforms, is used for the description of geological and geomorphological phenomena that need to be preserved and protected due to their significance or importance.

Geographical and Geological Setting

The study area is located in NW Greece (21° 52′ 32.48″ E and 40° 04′ 55.03″ N) at a mean altitude of 750 m above sea level. The site of the "Mpouharia" is characterized by a smooth relief surrounded by forests. Access is obtained by the existing road network, which connects the villages of Mirkovalto and Tranovalto. Other important geomorphological features in the broader area are the River Aliakmon to the west and its artificial lake (Lake Polyphytou), which lies to the north, as well as mountain Titaros to the northeast.

The geological formations of the "Mpouharia" area belong to the Pelagonian geotectonic unit, which is the westernmost zone of the Internal Hellenides (Brunn, 1956). The substrate of the broader area consists of both Pre-Alpine and Alpine formations such as gneisses, schists, limestones and flysch (Papanastasiou et al.,

A. Kelepertzis (✉)
National and Kapodistrian University of Athens, Faculty of Geology and Geoenvironment, Department of Economic Geology and Geochemistry, Greece
e-mail: kelepertsis@geol.uoa.gr

N. Evelpidou et al. (eds.), *Natural Heritage from East to West*,

DOI 10.1007/978-3-642-01577-9_5,

1998). These stratigraphic units form an internal basin (the Servia basin) which was created in the Pleistocene due to an episode of NW–SE extension (Pavlides and Mountrakis, 1987). The basin is located between 300 and 700 m above sea level and is flanked by mountain ranges up to 2,000 m, which are primarily composed of Mesozoic limestones, granites and Palaeozoic gneisses and schists (Mavridis and Kelepertzis, 1974; IGME, 1989). Continuous sedimentation resulted in the accumulation of a 600 m thick succession of Late Miocene to Early Pleistocene lake sediments with intercalated lignites and Quaternary deposits (Steenbrink et al., 2006). Quaternary deposits appear a reddish color and originate from the chemical and physical weathering of the metamorphic substrate (Desprairies and Faugères, 1971). From morphotectonic point of view, the wider area of Kozani-Grevena is dominated by a NE-SW striking normal fault zone (Hatzfeld et al., 1997; Pavlides et al., 1995; Meyer et al., 1996) which consists 3 distinct parts. The northern part of this zone defines the Servia fault, which is the most prominent feature of the area (Armijo et al., 1992).

Genesis, Evolution and Characteristics of Mpouharia and Nohtaria Landforms

The Mikrovaltos Mpouharia landforms are impressive, rare geological formations which resulted from thousands years of weathering that shaped columns of soil, sand and boulders topped by slate platforms. There are about 20 of these landforms, ranging from 3 to 6 m in height and 0.4 to 0.8 m in thickness, scattered among narrow watercourses and gorges on the sides of which are also found earthen stalagmites of heights of up to 10 m, ranging over 2 km. The Mpouharia (the word for "chimneys" in the local dialect) are also known by other names, such as giants, caryatids, fairies or demons, according to the prevailing legends surrounding their origins. They served as a refuge for local residents during invasions and wars (the War of Independence and occupation during World War II). Over recent years, the number of visitors to the Mpouharia has increased dramatically since tourism agencies organize regular excursions to the site. In the surrounding slopes of the wider area, there are also pyramid-shaped landforms which are called Nohtaria; these extend over an area of about 2 km long.

Mpouharia and Nohtaria landforms are unique in Greece, and so are of significant importance. Their formation is dated to 70,000 years ago. Similar landforms are present in Turkey in the region of Cappadocia (Aydan and Ulusay, 2003), but their characteristics differ because of the different primary and secondary hosting rocks. Mpouharia are sedimentary boles that were formed by the influence of weathering processes, during late geological periods. Their fragments consist of weathered debris, which originate from metamorphic rocks of low or medium cohesion. Water run-off during long periods either carried away these materials or cemented them forming the Mpouharia's main body, topped by a capping plate. The total number of the sedimentary boles is twenty; their height varies between 2 and 10 m.

As has been already mentioned, the upper geological sequence of the area consists of Neogene formations and recent Quaternary deposits. Most of the Quaternary deposits have a reddish color and originate from the chemical and physical corrosion of metamorphic rocks. Natural sections along the main torrent that crosses the Quaternary deposits revealed the existence of two distinct horizons: (a) the upper, thinner, one consisting of some fine grained, but mainly coarse grained, clastic materials and (b) the lower one of greater thickness, mainly characterized by fine grained clastic species. The Mpouharia landforms were formed by the natural weathering of the lower stratigraphic horizon, creating in that way long vertical sedimentary "boles" or "chimneys". On the other hand the pyramid-shaped formations, Nohtaria, which extend in an area, about 2 km in length, across the gorge, are attributed to the weathering of Quaternary deposits.

"Mpouharia" sedimentary boles

These processes took place after the formation of the drainage system. The top covers of Mpouharia consist of large plates of non-weathered metamorphic rocks (gneisses, schists) which remained above the long vertical sedimentary boles after the erosion of the upper stratigraphic horizon. These boles along with their top covers reflect the mature erosional stages and have been cemented during the diagenetic processing of the Quaternary deposits.

The distinction between these two horizons leads us to the conclusion that during the formation of the lower stratigraphic layer of greater thickness, climatic conditions were warm with elevated humidity, resulting to the natural weathering of gneisses and schists. The existence of coarse grained materials even at great distances reveals the direct impact of flood events. Furthermore, the gradation of the species that are hosted in cementing material of both Mpouharia and Nohtaria reflects their relative distance from the bedrock (gneisses and schists). As it can be assessed by the distribution of their size, Mpouharia contain coarser grained materials and their distance from the outcropping bedrock is in the range of a only few tens of meters. On the other hand, Nohtaria contain finer grained materials, and their distance from the outcropping formations of the substrate is greater (about 2 km). Probably this is the factor responsible for the different shape and specific characteristics of Mpouharia and Nohtaria.

The formation of local drainage system followed the deposition of the weathered materials over the older metamorphic rocks (gneisses and schists) and over older formations of the Neogene substrate such as lacustrine marls and clays. More specifically, active tectonism and weathering processes, due to water run-off, created a gorge that hosted the main stream, with several branches. The continuous weathering, along with the specific geological conditions, contributed to the formation of Mpouharia and Nohtaria landforms.

"Nohtaria" landforms

Unusual pyramid-shaped landforms called "Nohtaria"

Aeolian effects made a secondary contribution to their formation secondary. The enrichment of rainwater by soluble chemical materials, such as SiO_2, calcium, magnesium, potassium, sodium and iron, due to chemical reactions with the weathered materials of Quaternary deposits, proved to be determining factor for the cementation of the fine and coarse grained materials of Mpouharia. The cement materials consists quartz, opal, clay minerals, calcium carbonate and Fe oxides. In more details, X-Ray Diffraction analysis was performed in order to determine the mineralogical composition of the Mikrovaltos landform material. The results showed the existence of quartz, albite, microcline, illite and muschovite in descending order of concentrations. The above mineralogical composition reflects the impact of gneisses.

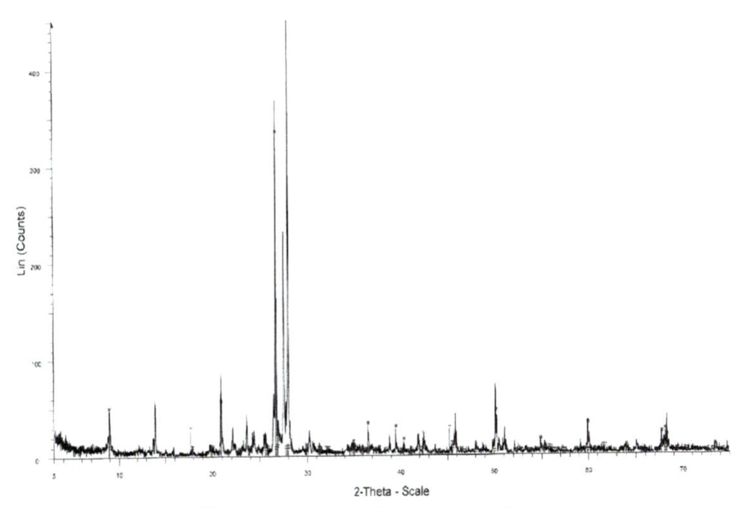

XRD analysis of the Mpouharia landform

As it can be assessed, during the weathering processes, gneisses and schists were broken down to smaller pieces of variable sizes mainly consisted of quartz and clay minerals. The above materials constitute a non cohesive weathered mantle which covers the gneisses and the older rocks. The main factors which contributed to the weathering of gneisses and schists are (a) climatic conditions and (b) specific geological characteristics.

"Mpouharia" rock pinnacles

In more details, the climatic conditions through the parameters of rainfall, temperature and humidity directly affected the weathering processes. Rainwater and subsequently run off, which was enriched in carbonic acid (H_2CO_3), often acted as a weak acid causing the degradation of gneiss' minerals, as it has been already mentioned. Variations of temperature caused mineral expansion or contraction, resulting to the formulation of thermoclasts. Furthermore, the schistosity and the phylloid structure of gneisses and schists which constitute their specific geological characteristics, favored the corrosion and finally the separation of their constituents.

Conclusions

Geological research over recent years has revealed a geosite of great importance in the Mikrovaltos area, of NW Macedonia (Greece). This site consists of impressive landforms resembling chimneys and pyramids, called Mpouharia and Nohtaria respectively. Their age of creation is estimated to be about 70,000 years, and their formation can be attributed to slow but continuous processes – the weathering of geological formations. The specific characteristics of these formations encouraged this process, which was favoured by the warm and humid climatic conditions. Rainwater, enriched in carbonic acid, acted as a weak acid and caused the mineral degradation of the substrate formations. The chemical action of rainwater, enriched in SiO_2, calcium, magnesium, potassium, sodium and iron, also led to the cementation of the fine and coarse grained materials of Mpouharia, and Nohtaria by cement materials such as quartz, opal, clay minerals, calcium carbonate and Fe oxides. Interest in these landforms has increased in recent years; several observers or tourists visit them daily. Hence it is important that an integrated management plan, concerning their protection, preservation and promotion, should be established.

References

Armijo, R., Lyon-Caen, H. and Papanastassiou, D. 1992. East-west extension and Holocene normal-fault scarps in the Hellenic arc. Geology 20, 491–494.

Aydan, O. and Ulusay, R. 2003. Geotechnical and geoenvironmental characteristics of man-made underground structures in Cappadocia, Turkey. Eng Geol 69, 245–272.

Brunn, J.H. 1956. Contribution à l'étude géologique du Pinde septentrional et d'une partie de la Macédoine occidentale. Ann Géol Pays Hell 7, 1–413.

Desprairies, A. and Faugères, L. 1971. Précisions stratigraphiques et sédimentologiques sur les dépôts néogènes du bassin de Servia (Macédoine occidentale, Grèce). Bull Soc Géol Fr 8(1–2), 67–84.

Hatzfeld, D., Karakostas, V., Ziazia, M., Selvaggi, G., Leborgne, S., Berge, C., Guiguet, R., Paul, A., Voidomatis, P., Diagourtas, D.. Kassaras, I., Koutsikos, I., Makropoulos, K., Azzara, R., Di Boma. M., Baccheschii, S., Bernard, P. and Papaioannou, C. 1997. The Kozani-Grevena (Greece) earthquake of 13 May 1995, revisited from a detailedSeismological study. Bull Seism Soc Am 87(2).

IGME. 1989. Seismotectonic map Of Greece, scale 1:250.000.

Mavridis, K. and Kelepertzis, A. 1974. Geological map of Greece, sheet Knidi, IGME

Meyer, B., Armijo, R., Massonnet, D., De Chabalier, J.B., Delacourt, C., Ruegg, J.C., Achache, J., Briole, P. and Papanastassiou. D. 1996 The 1995 Grevena, Northern Greece, earthquake. Fault model constrained with tectonic observations and SAR interferometry. Geophys. Res Lett 23(19), 2677–2680.

Papanastasiou, D., Drakatos, G., Voulgaris, N., and Stavrakakis, G. 1998. The May 13, 1995. Kozani-Grevena (NW Greece) earthquake: Source, study and its tectonic implications. J Geodynamics 26(2–4), 233–244.

Pavlides, S., Zouros, N., Chatzipetros, A., Kostopoulos, D. and Moudrakis, D. 1995. The 13 May 1995 western Macedonia. Greece (Kozani Grevena) earthquake; preliminary results. Terra Nova 7(5), 544–549.

Pavlides, S.B. and Mountrakis, D.M. 1987. Extensional tectonics of northwestern Macedonia, Greece, since the late Miocene. J Struct Geol 9(4), 385–392.

Steenbrink, J., Hilgen, F.J., Krijgsman, W., Wijbrans, J.R., Meulenkamp, J.E. 2006. Late Miocene to Early Pliocene depositional history of the intramontane Florina – Prolemais-Servia basin, NW Greece: Interplay between orbital forcing and tectonics. Paleogeography, Paleoclimatology, Paleoecology 238, pp. 151–178.

The Gorge of the Angitis River at "Stena Petras" Near the Alistrati Cave. A Magnificent Piece of Natural Architecture in Eastern Macedonia, Greece

Theodoros Astaras

Introduction of the Study Area

The Angitis river principally drains the Drama basin and also the former marshes of Philippi. It then flows in the spectacular gorge called "Stena Petras", just near the almost horizontal Alistrati caves and enters into the adjacent plain (basin) of Serres, where it joins the Strymon (Struma) river, discharging southwards to the Strymon gulf. The Angitis gorge and Alistrati caves are underlain by the Rhodope marbles. The outcrops of Alistrati marbles rise up to 273 m and are strongly jointed and fissured, mainly by NW-SE and NE-SW systems, with the NE-SW system being more dominant. Some faults oriented to NW-SE and NE-SW also occur in the marbles (Dimadis and Zachos, 1986).

The base-level of the Angitis gorge has an elevation of about 60 m at the entrance of the gorge and of about 30 m at its exit; with undulated hills to both sides, SSW and NNE of the gorge, reaching elevation between 120 and 160 m. The width and the depth of the gorge are 25–40 m and 50–70 m respectively. 250 m NNE of the gorge, the cave of Alistrati has developed (H.A.G.S, 1945, 1968, 1969; Symeonidis et al., 1977).

The objective of the study was to describe the geomorphic effect of ground (subterranean) water solution through the Alistrati Palaeozoic marbles of East Greek Macedonia, which has resulted, by subterranean meander cut-off, in the formation of the magnificent Angitis gorge with nearly vertical symmetric valley sides.

T. Astaras (✉)
Department of Physical and Environmental Geography, School of Geology, Aristotle University of Thessaloniki, University Campus 541 24 Thessaloniki-Greece
e-mail: astaras@geo.auth.gr

N. Evelpidou et al. (eds.), *Natural Heritage from East to West*,
DOI 10.1007/978-3-642-01577-9_6, © Springer-Verlag Berlin Heidelberg 2010

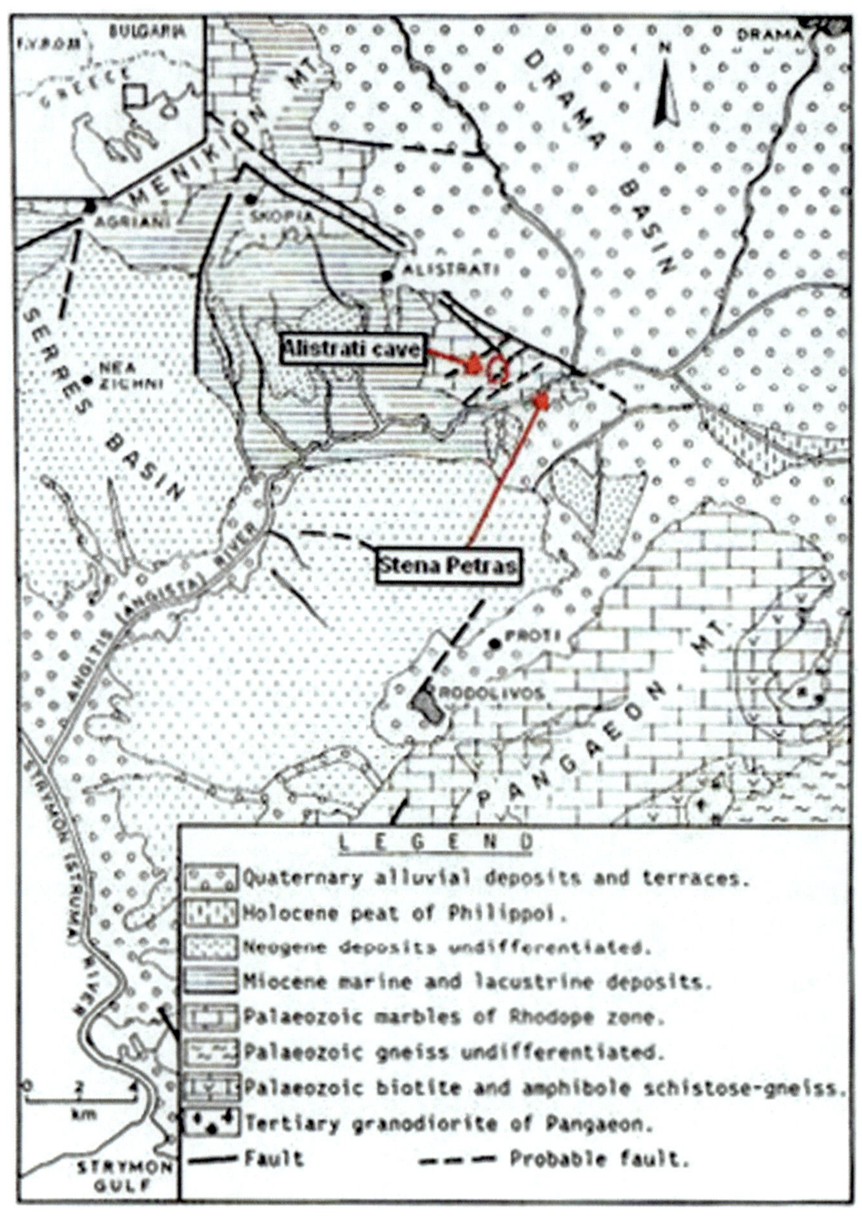

Geological map of the intermontane area between the Menikion and Pangeon mountains along the Angitis river (from Dimadis and Zachos, 1986 and Astaras 1988, modified)

The Formation of Angitis Gorge at Stena Petras

According to Astaras (1988), the Angitis gorge at Stena Petras is an "entrenched meander" which has been caused primarily by underground cut-off (karst self-piracy) of a meander stream that was flowing SE of the gorge, just outside the marble formation through the more easily eroded Neogene and Quaternary sediments, and secondarily by the accompanied fluvial incision within the marbles. This is in contrast to "ingrown meander" watercourses (Vavliakis et al., 1986), which usually result solely from fluvial processes (erosion, deposition, rejuvenation/incision) (Whittow, 1986). Most geomorphologists and geologists consider that deep Karstic ground erosion (solution) is more erosive than surface (fluvial) erosion, resulting in cave formation at the level of water-table, where ground-water flow is concentrated in its greatest volume (Rhoades and Sinacori 1941; Kaye, 1957; Bogli, 1964; Williams, 1969; Jennings et al., 1976; Palmer, 1982; Jennings, 1985).

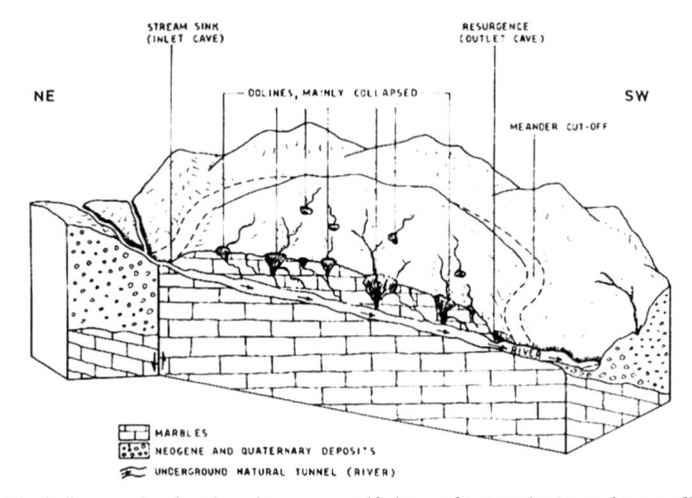

Block diagram showing the subterranean self piracy of a meander (meander cut-off through a meander spur) of the Angitis river. (From Astaras, 1988)

The underground cut-off (self-piracy) may has taken place along a natural karst tunnel that developed across the neck (waist) of the meander spur in the marbles, being controlled by strike, dip and unloading joints and other fissures in the marbles.

This subterranean cut-off through the meander spur consisted of three parts: (1) a swallow hole (stream-sink) on the upstream part of tha meander spur (near the village of Simvoli), into which water sank, (2) an underground route or tunnel, through which the water ran, (3) an outlet of the water downstream from the meander spur sink, 6–7 km SW of it. Therefore, the water was escaping from the river through the sink(s) and pierced the spur facing it.

The underground tunnel (cave) eventually was de-roofed by deep karstic ground erosion (the solutional and corrosive process of percolation water) and became a gorge with steep (nearly vertical) symmetric valley-sides, that is, an "entrenched meander" (Whittow, 1986).

The subterranean cut-off enlarged to the point where all the water of the Angitis river (draining the Drama plain) went through it, developing the karst tunnel which, as it was gradually de-roofed, created locally "natural bridges" before all the roof collapsed and disappeared completely. One natural bridge, on the left rim of Angitis gorge exists. After the creation of the Angitis gorge at "Stena Petras" the meander loop may have eventually been abandoned leaving alluvial gravels along its course that may have been buried beneath other surfase materials.

By making this underground cut-off, the Angitis river has shortened its length and increased its gradient; consequently it has now been rapidly rejuvenated (incising) within marbles.

View to SW of the spectacular Angitis gorge. The artificial pedestrian bridge in the middle has been created recently in order to connect the villages and the highway on the SE with the Allistrati cave entrance on the NW

The author based his conclusion on the following evidence, which resulted from interpretation of the 1:45,000 scale air photos and field work: (1) The continuation of certain drainage lines (mainly dry valleys) developed on the marbles, from one side of the gorge to the other. This matching of the drainage lines, in the marble on both sides of the gorge, is explained only by the already mentioned subterranean cut-off and the resulted collapse of the roof of the karst tunnel. (2) The form (shape) of the present small meander loops, along the segment of the Angitis river, flowing in the gorge, has not developed by chance, as usually happens in normal entrenched meanders derived from fluvial processes; rather, it has been controlled by the tectonic structure of the marbles (faults, joints etc.) (3) The meander-like cave of Alistrati, just 250–300 m to the west of the gorge, is oriented approximately parallel to it. (4) The existence of an arch of a "natural karst bridge" exactly on the rim of the Agitis gorge (half distance from the entrance of the gorge); this may be a part of the old cave network which later became the present gorge. (5) The incised meandering course of the Angitis gorge, which has steep (nearly vertical), symmetric valley-sides ("entrenched" meander), and an absence of fluvial terraces. This course indicates that the incision of Angitis river was relatively rapid, and not slow as is usually the case with "ingrown" meanders (which normally have asymmetric valley-sides, due to rejuvenation occurring more slowly than the development of meanders). Consequently, the rapid, vertical and symmetric erosion of the channel floor of the Angitis gorge (an "entrenched meander") is explained more clearly by subterranean meander cut-off and subsequent underground tunnel (cave), which eventually has been de-roofed completely and formed a gorge, than by slow fluvial processes along an "ingrown" meander. (Astaras, 1988).

The formation of the Angitis gorge (by subterranean cut-off) took place long ago (Upper Pleistocene), possibly during the period of high flood events and lower sea level, when the capacity of the underground system was exceeded (Astaras, 1982).

Use of the Angitis Gorge as a Recreation and Education Area

Both the spectacular Angitis gorge and the adjacent cave of Alistrati are the best geological monuments in Eastern Macedonia, Greece. In particular the gorge of Angitis at "Stena Petras" is a magnificent piece of natural architecture.

Close-up of the "Natural bridge" exactly on the left rim of Angitis gorge, flowing SW. View from the right rim of Angitis gorge

During recent years, preservation and land reclamation of these geological monuments have been started; and they are now in an appropriate condition and open to the public (foreign and domestic), as an educational and recreation area in Northern Greece.

Panoramic view taken from the right side of the Angitis gorge

There is a pedestrian bridge which has been installed to link suitable points, where the opposite rims of the Angitis gorge are smooth, thereby connecting the villages on the SE with the cave entrance on the NW.

Also, along the Angitis gorge, on both sides, pedestrian paths have been created connecting various kiosks, which are used as resting points for tourists.

After the above preservation and reclamation works, many tourists and students, High school and University, both domestic and foreign, visit the area during the whole year.

View from the NE rim of the gorge of the Angitis, showing pedestrian paths along the gorge in the front part of the picture and a kiosk and path to the Alistrati cave entrance in the middle and background of the picture

References

Astaras, T. 1982–1983. On the stream capture between the tributaries "Megalo Potami" and "Xiropotamos" of the Gallikos river. Geomorphological significance of stream captures to the study of fluvial placers and in general to landscape evolution of a certain area. *Ann. Geol. Pays Hellen.*, 31: 725–740, Athens (in Greek with English summary).

Astaras, T. 1988. A Karst Stream subterranean "autopiracy" of Angitis river flowing in the gorge of Stena Petras, near Alistraty, East Macedonia, Greece: A contribution to the evolution of the epigenetic valley of Angitis river. *Ann. Geol. Pays Hellen.*, 33: 463–473.

Bogli, A. 1964. Mischungskorrosion-ein Beitrag zum Verkastrungsproblem; Erdkunde, 18: 83–92.

Dimadis, E. & Zachos, S. 1986. Geological map of Rhopope massif, Scale 1:200.000. Institute of Geology and Mineral Exploration (I.G.M.E.), Xanthi branch.

Hellenic Army Geographical Service (H.A.G.S.) 1945, 1968, 1969. Topographic maps of Prosotsani and Drama, scale 1:50.000 (1969); and airphotos of approx. scale 1:45.000 (1945) and 1:55.000 (1968) Athens.

Jennings, J., Brusch. J., Nicoll, S. & Spate, A. 1976. Karst stream self capture at London Bridge, Bura Creek, N.S.W., *Aust. Geogr.*, 13: 238–249.

Jennings, J. 1985. *Karst Geomorphology*. Basil Blackwell, Oxford, UK: 88–94.

Kaye, C. 1957. The effect of solvent motion on limestone solution. *Journ. Geol.*, 65: 35–46.

Palmer, A. 1982. Geomorphic interpretation of karst features, Summary proceedings: 9. Thirtheent Annual Geomorphological Symposium (USA) (A continuation of SUNY Binghampton Geomorphology Symposium series), entitled "Groundwater as a Geomorphic agent" Troy, N.Y. (USA).

Rhoades, R. & Sinacori, M. 1941. Pattern of ground-water flow and solution. *Journ. Geol.*, 49: 785–94.

Symeonidis, N., Dilaras, G., Tsimbanis, E., Papadopoulos G. & Constantacatos E. 1977. The cave of Alistrati in Serres (Greece). *Bull. Greek Speleol. Soc.*, 14: 64–81, Athens.

Vavliakis, E., Psilovikos, A. & Sotiriadis, L. 1985. The epigenetic valley of Angitis river in relation with the evolution of the basins in Serres and Drama. I.G.M.E., Geol. and Geoph. Research, Special issue, 1986: 5–14. Athens.

Whittow, J. 1986. *Dictionary of Physical Geography*. Penguin Books Ltd, England: 592.

Williams, P. 1969. The geomorphic effects of ground water. In Chorley R. (ed.), *Water, Earth and Man*, London, Methuen & Co. Ltd: 269–284.

Volcanoes "Monuments of Nature"

Konstantinos G. Kyriakopoulos

Introduction

Volcanoes are the surface expression of the interior part of the earth's crust. In general they are formed when melting material rises from depth and leaks into the earth's crust. The incandescent (molten) rock underlying the volcano, called magma, erupts as lava and pyroclastic products on the surface, forming a large cone. Volcanoes occur into four main contexts: (i) subduction (convergent) zones, (ii) areas of spreading (iii) intraplate geotectonic environments and (iv) at hot spots in the lithosphere. As monuments of nature, volcanoes always have cultural dimensions. They are very attractive places of great interest for geology as well as for human history and environmental evolution. In the past the observation of, and reference to, natural phenomena has been in poetry. Throughout the world, volcanoes, rocks, mountains, lakes, caves or other geological formations have been used for rituals, or inspired artists, travellers and myths. For example, the description of the eruption, in AD 79, of Vesuvius by Pliny the Younger is essential for volcanology. Volcanoes are attractive places for anyone wishing to visit, explore, and photograph; they offer a fascinating adventure for professional volcanologists as well as for amateur enthusiasts.

Cultural Landscapes: Effects and Forms of Volcanic Activity

The primary effects of volcanic activity include pyroclastic deposits, lava flows, ash flows (hot avalanches, nuees ardentes and ignimbrites), lateral blast, ash falls and gases. Mudflows, pyroclastic flows and ash falls pose the greatest danger to property and people in regions of active volcanism. In active tectonic regions, where the volcanic layered rocks give rise to unstable slopes, landslides are common and easily

K.G. Kyriakopoulos (✉)
Section of Mineralogy and Petrology, Department of Geology and Geoenvironment, National and Kapodistrian University of Athens, Panepistimioupolis, Gr-15784, Athens, Greece
e-mail: ckiriako@geol.uoa.gr

N. Evelpidou et al. (eds.), *Natural Heritage from East to West*,
DOI 10.1007/978-3-642-01577-9_7, © Springer-Verlag Berlin Heidelberg 2010

modify the morphology of the area. Volcanic landscapes contain diverse landforms. The most recognizable of these are volcano edifices, calderas and lava domes. Each of these landforms can vary remarkably in size, shape, composition and eruptive history.

Volcanic landforms morphology are the end result of two opposing forces. Constructive forces build up volcanoes during their ongoing volcanic activity. Destructive forces, such as weathering and erosion, tend to destroy the structures of a volcano edifice and in many cases forming calderas depression.

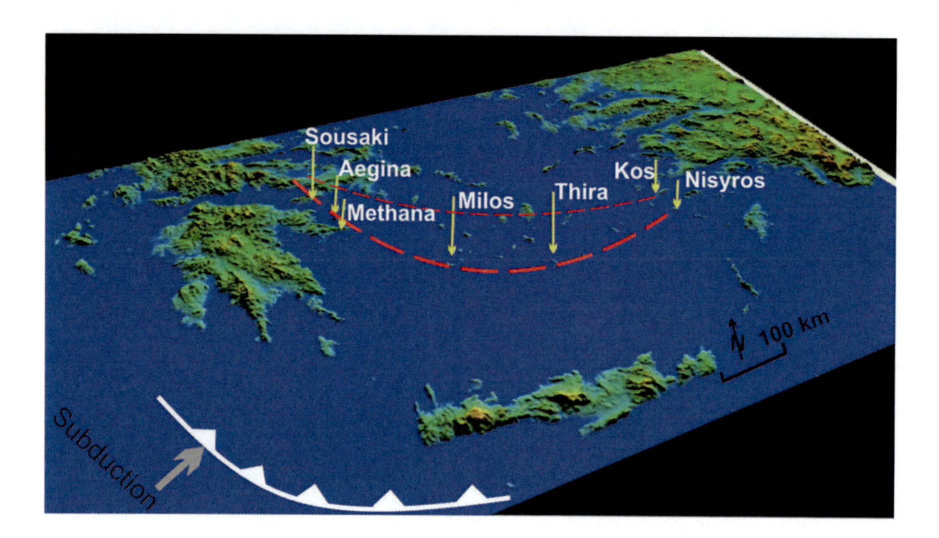

Volcanoes in the Aegean Region

Volcanic activity in the south Aegean area took place during the Plio-Quaternary period. The present Aegean geodynamic model reflects an active convergent zone, where the continental micro-plates exhibit a complex interaction under the influence of overall N-S convergence between the African plate and the Eurasian plate (Smith, 1971; Ninkovich and Hays, 1972; Angelier et al., 1977; Dewey and Sengor, 1979; Keller, 1982). The Aegean volcanic arc lies ca. 150 km above the subduction zone, which is the form of an amphitheatre-like conical surface (a spoon form), reaching a maximum depth of 190 km in the central part of the South Aegean Volcanic Arc (Spacman al., 1988; Wortyel and Spakman, 1992; Papadopoulos et al., 1986; Papazachos, 1990). The volcanic products of Southern Aegean region, of a calc-alcaline character, developed at various volcanic centers from Soussaki to Nisyros via Methana-Aegina-Poros, Milos and Santorini. The volcanism started 3.5 Ma ago and still continues in the form of post-volcanic activity (Nickolls, 1971; Keller, 1982; Fytikas et al., 1984, 1986; Mitropoulos and Magganas, 1988; Mitropoulos and Katerinopoulos, 1993).

The main landscape evidence in the volcanic areas of the south Aegean are:

1. the shape and development of the volcanic structures,
2. the morphology and evolution of the main eruptive event of a volcano in space and time,
3. the petrological and mineralogical composition as well as the texture and the microstructure characteristics of the volcanic products,
4. post volcanic phenomena (gas emission, thermal springs, geothermal fields) connected to the local tectonic setting and to a general geotectonic regime affecting the whole area.

Sousaki Volcano

The Sousaki area is located about 65 km west from Athens, close to the Isthmus of Corinth. It represents the NW end of the active Aegean volcanic arc. Here, sparse outcrops of dacite rocks are the remnants of late-Pliocene to Quaternary volcanic activity ranging between 4.0 and 2.3 Ma (Pe-Piper and Hatzipanagiotou, 1997). Drilling exploration assessed the presence of a low enthalpy geothermal field, revealing two permeable formations. The first is located at a shallow depth (< 200 m) and the second at a deeper level (500–1,100 m). All geothermal waters are of Na-Cl type and have temperatures in the range 50–80°C and salinities in the range 39–49 g/l (Fytikas et al., 1995; D' Alessandro, et al., 2006). Geodynamically, the whole region between Corinth and the Gulf of Saronikos is very active with frequent earthquakes; by implication these are controlled by active tectonic mechanisms. The post-volcanic activity still observed today is mainly manifested by emanation of warm fluids, while widespread fumarolic alteration and warm (35–45°C) gas emissions are still recognizable. The reaction of these gas phases with the surrounding rocks forms secondary mineral aggregates, (Kyriakopoulos et al., 1990; Stiros, 1995; Kelepertsis et al., 2001).

Landscape of the Sousaki volcano and secondary mineral aggregates due to H_2S emanation

Sousaki main volcanic gases emission (CO_2 as the main component and CH_4 and H_2S as minor components)

Sousaki volcano, secondary minerals due the reaction of emitted gases with surrounding rocks

Methana-Aegina-Poros Volcanic Group

Methana, a peninsula about 44 km^2 in size, located on the north-eastern coast of Peloponnesus, is the westernmost active volcanic system of the South Aegean volcanic arc. (Pe-Piper and Piper, 2002; Dietrich and Gaitanakis, 1995). The volcanic sequences of the peninsula principally consist of andesite and dacite lava domes and flows extending radially from its central part. Volcanic activity probably started at the transition of Plio-Pleistocene. However, the oldest dated rocks give ages of about 0.9 Ma. The most recent volcanic activity was a flank eruption that produced andesitic lavas at Kammeno Vouno around 230 BC; this was described by Strabo (Pe-Piper and Piper, 2002; D' Alessandro et al., 2007).

Volcanic dome at the north western part of Methana peninsula

Methana peninsula volcano, main crater at kameni Chora

About 2/3 of the whole area of the island of Aegina is composed of volcanic rocks. The volcanic activity evolved into two distinctive cycles (4.7 and 4.3 Ma)

and (3.9 and 3.0 Ma). The volcanic products of the first cycle range from mainly rhyodacitic pumice flows and tuffites to andesitic pillow lavas and hyaloclastites. The volcanic rocks of the second cycle cover the central and south part of the island, in the Oros mountain (Muler et al., 1979; Dietrich et al., 1993; Matsuda et al., 1999). Post volcanic activity is manifested by a hot spring, temperature about 25°C, and a low enthalphy geothermal system (30–40°C).

On Poros volcanic rocks, occurring in a small area of about 1 km^2, are mainly composed of lava flows and a lava dome, about 3.11–2.6 Main age (Pe, 1975; Fytikas et al., 1986; Mitropoulos et al., 1987; Matsuda et al., 1999).

Methana volcano, historic eruption lava flow (230 BC) at Kameni Chora

Thermal spring in the Methana town

Volcanic andesitic dome at Aegina island of columnar form

Milos-Kimolos Volcanic Field

The island of Milos (~150 km^2) together with Kimolos, Polyegos, Antimilos, Prasonisi and Glaronisia constitute the beautiful Milos volcanic complex. Is situated in the south-western part of the Attico-Cycladic Massif in the South Aegean volcanic Arc (Le Pichon and Angelier, 1979; Innocenti et al., 1981; Mitropoulos et al., 1987, 1992). The volcanic products of Milos consist of Plio-Pleistocene age rocks, mainly of rhyolitic-rhyodacitic and andesitic composition, which show a typical calcalkaline affinity. The erupted products are mainly acidic tuffs and pumice, pyroclastic flow and lahar deposits, while during extrusive activity andesitic domes and in places lava flows are formed. The volcanic rocks were erupted between 3.5 and 0.08 Ma ago. The last eruption took place in the area of the Tsingrado volcano more than 90 Ka ago.

Fyriplaka volcano edifice at south part of Milos island

Sarakiniko volcanic landscape, north pert of Milos island

The present topography of Milos is the result of both volcanic and tectonic activities. The multicolour volcanic formations, together with their hydrothermal activity products, constitutes the variety of shapes like the beautiful "Sarakiniko" area. The younger volcanic products are most easily recognized, as lava flows, domes and tuffs. According to abundant evidence, the area of Milos and Kimolos, and the central and the eastern part of the South Aegean volcanic Arc, has been affected by the strong extensional tectonic regime, providing favorable conditions for the magma to rise and form vast magmatic chambers near the surface (Kelepertzis and Kyriakopoulos, 1991; Kyriakopoulos, 1998; Fytikas et al., 1986, 1989).

Obsidian outcrop at west Adamas village Milos island

Native sulphur, secondary formation, at Kalamos area Fyriplaka volcano, Milos island

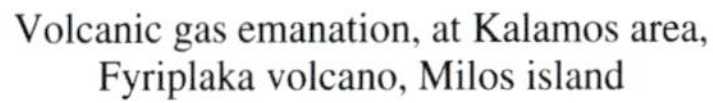

Volcanic gas emanation, at Kalamos area, Fyriplaka volcano, Milos island

Natural volcanic work of art, named "Skiadi" in the central part of Kimolos island

Santorini Volcanic Group

The Santorini volcanic group is composed of Thira, Thirasia, Christiana, Palea and Nea Kameni islands and the Columbo submarine volcano. It is found in the central part of the south Aegean active volcanic arc. The eruption of Santorini was a violent caldera-forming event that ocurred in the last 100 Ka. The beautiful caldera, 13 km wide, is bounded by cliffs up to 300 m high and extends about 450 m below sea level (Druitt et al., 1989; Fytikas et al., 1990; Barberi and Carapezza, 1994). The volcanic extrusive, effusive and, to a minor extent, explosive activity produced lava flows, domes and pyroclastic sequences; these formed Palea and Nea Kameni islands between 197 BC and 1950 AD (Fouque, 1879; Washington, 1926; Ktenas, 1927; Georgalas, 1953). Lavas from the Akrotiri, Micro Profitis Ilias, Megalo Vouno and Skaros volcanic centers span the age of the whole volcanic field and range in composition from basalt to rhyodacite. Petrochemical data show that the volcanic rocks result from fractional crystallization, combined with assimilation and magma mixing, in intra-crustal magma chambers located at depths of approximately 7–14 km (Druit et al., 1989; Mitropoulos and Tarney, 1992; Skarpelis et al., 1992; Vougioukalakis et al., 1995; Zelmer et al., 2000).

Most of the active volcanoes of the Mediterranean region are characterized by mainly explosive activity and therefore represent very hazardous areas for man. The greatest hazard for the Santorini area is indicated by the effects of the caldera forming Minoan eruption of the late Bronze Age (about 3.6 ka BP). During the twelve most explosive eruptions, about 30 cubic kilometres of incandescent magma were erupted (Druit et al., 1989). At Milos and Nisyros no magmatic events occurred in historical times. The islands of Santorini, Milos and Nisyros with small permanent populations during winter show drastically population increase during the tourist seasons, which considerably enhances the volcanic risk (Kyriakopoulos et al., 2003).

Santorini strato-volcano, caldera view near Phira village

Santorini volcano, landforms of pyroclastic deposits

Santorini volcano, sunset at Nea and Palea-Kameni volcanic islands

Santorini volcano, weathering and erosion pumice material of the 1650 B.C. "Minoan eruption"

Santorini volcano, uppermost layers of pumice deposits of the 1650 B.C. "Minoan eruption"

Nisyros – Yali and Kos Volcanoes

Nisyros stratovolcano is situated at the eastern end of the south Aegean volcanic arc. The oldest age of the volcanic rocks is about 160 Ka and the most recent phreatic eruption took place in 1867 AD. The morphology and shape of the volcano is controlled by a caldera depression of 4 km diameter. The exposed stratigraphy starts with pillowed basaltic andesite and pillow breccia and evolves towards more felsic volcanism. Post-caldera activity, as rhyodacitic domes, rises above the caldera rim to a maximum high point of St. Helias about 700 m. The recent caldera-forming eruption created a tephra dated 46,000 Ka BP. The more recent and intense explosions of the volcano in historical times (1422, 1771, 1873, 1887, 1888) are hydrothermal and created the 10 small craters: Mikros Polivotis, Alexandros or Flegethron, Polivotis, Ahileas, Megalos Polivotis, Logothetis, Mikros Stefanos, Stefanos and Kaminakia, on the caldera floor. Stefanos crater is the largest, with a depth of 27 m and diameter of 330 m (Martelli, 1917; Di Paola, 1974; Keller, 1980; Papanikolaou et al., 1991; Francalanci et al., 1995; Seymour, 1996; Vougioukalakis et al., 1998; Kyriakopoulos et al., 2003; Margari, 2004; D' Alessandro et al., 2008).

Nisyros volcano, hydrothermal crater (Stefanos) on caldera floor

Nisyros volcano, weathering and aeolian erosion phenomena on pyroclastic deposits

Yali volcano,obsidian outcrop at the central part of the island

Kos tuff plateau deposits around Kefalos bay, south Kos island

Volcanism in South Kos island has formed different domes of dacitic and rhyolitic composition from 3.4 to 1.6 Ma. The last volcanic explosion (Kefalos area) of 161 Ka BP is an excellent example of large-scale caldera-forming eruption. This was the most catastrophic volcanic event in the Quaternary Hellenic Volcanic Arc. The Kos plateau Tuff of pumice flows covers an area of about 5,000 Km^2 (Bellon et al., 1979; Keller, 1982; Dalambakis, 1986; Keller et al., 1990; Smith et al., 1996; Allen and Cas, 1998; Palladino et al., 2008).

References

Allen S.R. and Cas R.A.F., (1998). Rhyolitic fallout and density current deposits from a phreatoplinian eruption in the eastern Aegean Sea. J. Volcanol. Geotherm. Res., 86: 219–251.

Angelier J., Cantagrel M. and Vilminot C., (1977). Neotectonique cassante et volcanisme plio-quaternaire dans l' arc eggeen interne: l' Ile de Milos (Grece). Bull. Soc. Geol. France, 19: 119–121.

Bellon H., Jarrige J.J. and Sorel D., (1979). Les activites magmatiques egeennes del' Oligocene a nos jours et leurs cadres geodynamiques.Donnees nouvelles et synthese. Rev. Geol. Dynam. Geogr. Phys., 21: 41–55.

Barberi F., and Carapezza, M.I., (1994). Helium and CO2 soil gas emissions from Santorini (Greece). Bull. Volcanol., 56:335–342.

D' Alessandro W., Brusca L., Kyriakopoulos K., Rotolo S., Michas G., Minio M. and Papadakis G., (2006). Diffuse and focused carbon dioxide and methane emissions from the Sousaki geothermal system, Greece, Geophy. Res. Lett., 33 LO5307.

D' Alessandro W., Brusca L., Kyriakopoulos K., Margaritopoulos M., Michas G., Papadakis G., (2007). Fluid geochemistry investigations on the volcanic system of Methana. Bullet. Geol. Soc. Greece. Vol. XXXVII, Proceedings of the 11th international Congress, Athens.

D' Alessandro W., Brusca L., Kyriakopoulos K., Michas G., Papadakis G., (2008a). Methana the westernmost active volcanic system of the south Aegean arc (Greece). J. Volcanol. Geoth. Res., 178: 818–828.

D' Alessandro W., Brusca L., Kyriakopoulos K., Michas G., Papadakis G., (2008b). Hydrogen sulphide as a natural air contaminant in volcanic/geothermal areas: the case of Sousaki, Corinthia (Greece). Environmental Geology. DOI 10.1007s00254-008-1453-3.

Dietrich V.J., Gaitanakis P., Mergoli I., and Oberhaensli R., (1993). Geological map of Greece, Aegina Island, 1:25000. Sixth Congress, Bull. Geol. Soc. Greece, Vol. XXVIII/3, pp. 555–566.

Di Paola G.M., (1974). Volcanology and petrology of Nisyros island (Dodecanese, Greece). Bull. Volcanol., 38: 944–987.

Dietrich V. and Gaitanakis P., (1995). Geological map of Methana peninsula (Greece). ETH Zurich, Switzerland.

Druitt T.H., Mellors R.A., Pyle D.M. and Sparks R.S.J., (1989). Explosive volcanism on Santorini, Greece. Geol. Mag., 126: 95–126.

Dewey J.F. and Sengor A.M.C., (1979). Aegean and surrounding regions: complex multiplate and continuum tectonicsin convergent zone. Bull. Geol. Soc. Am. 90: 84–92.

Fouque F., (1879). Santorin et seseruptions.Masson et cie, Paris.

Francalanci L., Varekamp J.C., Vougioukalakis G., Defant M.J., Innocenti F., Manetti P., (1995). Crystal retention, fractionation and crustal assimilation in a convecting magma chamber, Nisyros Volcano, Greece. Bull. Volcanol., 56: 601–620.

Fytikas M., Innocenti F., Manetti P., Mazzuoli R., Peccerillo A and Villari L., (1984). Tertiary to Quaternary evolution of volcanism in the Aegean region. J. Geol. Soc. London Spec. Publ. 17, 687–699.

Fytikas M., Innocenti F., Kolios N., Manetti P., Mazzuoli R., Poli G., Rita F., and Villari L., (1986). Volcanology and petrology of volcanic products from the island of Milos and neighbouring islets. J. Volcan. Geoth. Res., 28, 297–317.

Fytikas M., (1989). Updating of the geological and geothermal research on Milos Island. Geothermics, 18(4): 485–496.

Fytikas M., Kolios N. and Vougioukalakis G.E., (1990). Post-Minoan volcanic activity of the Santorini volcano. Volcanic hazard and risk. Forecasting possibilities. In hardy D.A., Keller J., Galanopoulos V.P., Flemming N.C., Druitt T.H. (eds.). Thera and the Aegean World III. The Thera Foundation, London, vol 2, 183–198.

Fytikas M., and Vougioukalakis G., (1995). Volcanic hazard in the Aegean Islands. In: T. Horlick-Jones, A. Amendola and R. Casale, Natural Risk and Civil Protection. E & FN Spon, Brussels, pp. 117–130.

Georgalas G., (1953). L' eruption du volcan de Santorin en 1950. Bull. Volcanol., 13:39–55.

Innocenti F., Manetti P., Peccerillo A. and Poli G., (1981). South Aegean volcanic arc: geochemical variations and geotectonic implications. Bull. Volcanol., 44:377–391.

Keller J., (1980). Prehistoric pumice tephra on Aegean islands.-C. Doumas (ed.) 'Thera and Aegean World II', vol. 2 London, pp 49–56.

Keller J., (1982). Mediterranean island arcs. In: Thorpe, R.S. (ed.). Andesites J. Wiley and Sons, New York, 307–326.

Keller J., Rehren T. and Stadlbauer E., (1990). Explosive volcanism in the Hellenic Arc: A summary and review. In: Hardly, D.A., Keller, J., Galanopoulos, V. P., Flemming, N. C., Druitt, T. H. (eds), Thera and the Aegean World III. The Thera Foundation, London, vol. 2, pp. 13–26.

Ktenas K., (1927). L' eruption du volcan des Kammenis (Santorin) en 1925, II. Bull. Volcanol., 4: 7–46.

Kyriakopoulos K., Kanakis-Sotiriou R., Stamatakis M., (1990). The authigenic minerals formed from volcanic emanations at Soussaki, West Attica peninsula, Greece, Can. Miner., 28: 363–368.

Kyriakopoulos K., (1998). K/Ar and Rb/Sr isotopic data of Micas from Milos island geothermal field. Ann. Geol. Pays. Hell., 38: 37–48.

Kyriakopoulos K., Vavassis, I., Brombach,T. and De Astis G., (2003). Active volcanoes in the Mediterranean region: Volcanic Hazards and Risk Management. The South Aegean Active volcanic Arc, Recent knowledge and future prospectives, The SAAVA Milos conference, book of Abstracts vol. p. 67.

Kelepertsis A., Alexakis D. and Kita I., (2001). Environmental geochemistry of soils and waters of Susaki area, Korinthos, Greece, Environ. Geochem. Health, 23: 117–135.

Kelepertzis A. and Kyriakopoulos K. (1991). Mineralogy and geochemistry of the Mn-mineralization from Vani area of Milos island-its genesis problem. Prakt. Acad. of Athens, vol. 66, pp. 107–121.

Le Pichon X. and Angelier J., (1979). The Helenic arc and trench system. Tectonophysics, 60:1–42.

Margari V., (2004). Late Pleistocene vegetational and environmental changes on Lesvos Island, Greece. Ph.D. 663 Thesis, University of Cambridge, Cambridge, UK.

Martelli A., (1917). Il gruppo eruttivo di Nisyros nel mare Egeo. Mem. Soc. Geol. Ital. Sc. (detta dei XL), Serie 3a, vol. XX.

Matsuda J., Senoh K., Maruoka T., Sato H. and Mitropoulos P., (1999). K-Ar ages of the Aegean volcanic rocks and their impications for the arc-trench system. Geochem. J., 33: 369–377.

Mitropoulos P., Tarney J., Saunders D. and Marsh N., (1987). Petrogenesis of cenozoic volcanic rocks from the Aegean Island Arc. J. Volcanol. Geoth. Res., 32: 177–193.

Mitropoulos P. and Magganas, A., (1988). Interisland variations of the oxygen fugasity (fO2) implied from Fe-Ti Mineral oxides chemistry in the Aegean volcanic Arc. Ann. Geol. Des Pays Hellen., 33: 147–149.

Mitropoulos P.and Tarney J., (1992) Significance of mineral composition variations in the Aegean island arc. J. volcanol. Geoth. Res., 51: 283–303.

Mitropoulos P. and Katerinopoulos A., (1993). Geochemical characteristics of volcanic rocks from the Eastern Aegean on an axis perpendicular to the volcanic arc. Bull. Geol. Soc. Geece, XXVIII/2, 209–220.

Muler P., Kreutzer H. and Harre W., (1979). Radiometric dating of two extrusives from a Lower pliocene marine section on Aegina Island, Greece. Newslett. Stratigr., 8: 70–78.

Nickolls A., (1971). Petrology of Santorini volcano, Cyclades, Greece. J. Petrol. 12: 67–119.

Ninkovich D. and Hays J.D., (1972). Mediterranean Island Arcs and origin of High potash volcanics. Earth Planet. Sci. Lett., 16: 331–345.

Palladino D., Simei S., Kyriakopoulos K., (2008). On magma fragmentation by conduit shear strees: Evidence from the Kos Plateau Tuff, Aegean Volcanic Arc. J. Volcanol. Geoth. Res., 178: 807–817.

Papadopoulos B.A., Kondopoulou D.P., Leventakis G.A., Pavlides S.B., (1986). Seismotectonics of the Aegean Region. Tectonophysics, 124: 67–84.

Papanikolaou D., Lekkas E., sakelariou D., (1991). Volcanic stratigraphy and evolution of the Nisyros volcano. Bull. Geol. Soc. Greece, XXV: 405–419.

Papazachos B.C., (1990). Seismisity of the Aegean and surrounding area. Tectonophysics, 178: 287–308.

Pe G.G., (1975). Strontium isotope ratios in volcanic rocks from the northwestern part of the Hellenic Arc. Chem. Geol., 15: 53–60.

Pe-Piper G. and Hatzipanagiotou K., (1997). The Pliocene volcanic rocks of Crommyonia, western Greece and their implications for the early evolution of the South Aegean arc, Geol. Mag., 134: 55–66.

Pe-Piper G. and Piper D J W., (2002). The igneous rocks of Greece. Borntraeger, Stuttgart, 645 pp.

Skarpelis N., Kyriakopoulos K. and Villa I., (1992). Occurrence and 40Ar/39Ar dating of a granite in Thera (Santorini, Greece). Geologische Rundschau, 81/3: 729–735.

Smith P.E., York D., Chen Y., Evensen N.M., (1996). Single crystal 40Ar/39Ar dating of a Late Quaternary paroxysm on Kos, Greece: Concordance of terrestrial and marine ages. Geophys. Res. Lett., 23: 3047–3050.

St. Seymour K., (1996). Geochemistry of the Yali volcano rhyolites and their relationship to the volcanic products of Nisyros, Aegean volcanic arc. N. Jb. Miner. Mh., H.2, 57–72.

Smith A.G., (1971). Alpine deformation and orogenic areas of the Tethys, Mediterranean and Atlantic. Geol. Soc. Amer. Bull., 82: 2039–2070.

Spacman W., Wortel M.J.R. and Vlaar N.J., (1988). The Hellenic subduction zone: a tomographic image and its geodynamic implications. Geophys. Res. Lett., 15(1): 60–63.

Stiros S.C., (1995). The 1953 seismic surface fault: implications for the modeling of Sousaki (Corinth area, Greece) geothermal field. J. Geodyn., 20: 167–180.

Vougioukalakis G., Francalanci L., Sbrana A., Mitropoulos. D., (1995). The 1649-1650 Kolumbo submarine volcano activity, santorini. Greece. In: F. Barberi, R. casale and M Fratta (eds.), "The European laboratory Volcanoes, Workshop Proceeding" European Commission, European Science Fondation, Luxembourg, pp. 189–192.

Vougioukalakis G., Sachpazi, M., Perissoratis C. and Lyberopoulou Th., (1998). The 1995-1997 seismic crisis and ground deformation on Nisyros volcano, Greece: a volcanic unrest?. 6th Int. Meeting on Colima volcano, Abstracts Vol.

Wortyel M.J.R.and Spacman W., (1992). Structure and dynamic of subducted lithosphere in the Mediterranean region. Proc. Kon. Ned. Acad. v. Wetensch., 95(3): 325–347.

Washington H S., (1926). Santorini eruption of 1925. Bull. Geol. Soc. Am., 37: 349–384.

Zelmer G., Turner S., Hawkesworth, C., (2000). Time scales of destructive plate margin magmatism: new insights from santorini, Aegean volcanic arc. Earth Planet. Sci. Lett., 174: 265–281.

Amber in Romania

Antonela Neacsu

Introduction

Amber is a fosssil resin originating from different types of conifers and certain flowering trees. It was formed worldwide at least 40 million years ago.

From the chemical point of view, amber is a high molecular weight cross-linked polymer, the product of esterification of the co-polymer of communal and communic acid with succinic acid, the latter being a degradation product of abietic acid (Rottländer, 1970). In addition to the polymeric material, amber is also composed of lower molecular weight volatile compounds (2–5%) including aromatic hydrocarbons (cymenes) and monoterpenes (borneol, camphor, fenchyl alcohol and fenchone). The empirical formula of Baltic amber has been given both as $C_{10}H_{16}O$ (Brydson, 1999) and $C_{79}H_{10.5}O_{10.5}$ (Frondel, 1968) (in Shashoua, 2002). Mineralogically, amber is a *mineraloid*.

Species and Varieties

There are many amber species and varieties; the best known species are: Baltic amber or *succinite*, Dominican amber, *simetite* – Sicilian amber, *burmite* – a resin extracted in the superior Burma and processed in China, amber from Borneo, amber from Alava (Spain) and *romanite* – amber from the Romanian Carpathians.

A. Neacsu (✉)
University of Bucharest, Romania
e-mail: antonela@geo.edu.ro

N. Evelpidou et al. (eds.), *Natural Heritage from East to West*,
DOI 10.1007/978-3-642-01577-9_8, © Springer-Verlag Berlin Heidelberg 2010

Yellow succinite, Helgoland, Germany - Museum of Mineralogy, Department of Mineralogy, University of Bucharest

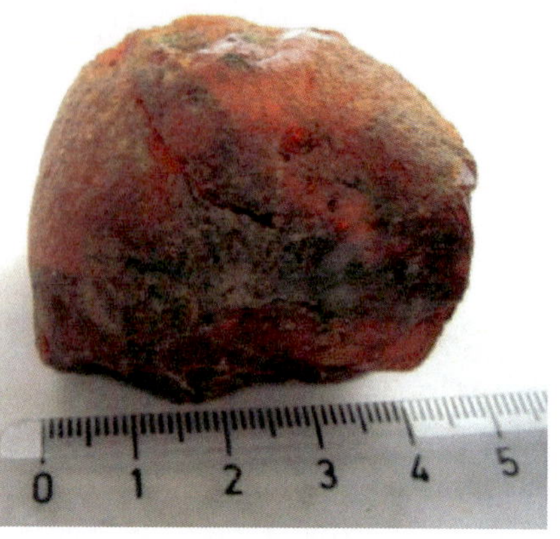

Succinite, Samland Peninsula, Russia - Museum of Mineralogy, Department of Mineralogy, University of Bucharest

Succinite, Poland - Museum of Mineralogy, Department of Mineralogy, University of Bucharest

In 1891, Otto Helm noticed that *"Rumänite"* is a variety of Buzau amber with a clear yellow, blackish, brown or reddish coloration due to alteration; it is often burnt (over-oxidized and friable).

Resinosis – an Uncommon Phenomenon in the Geological Past

An unusually large secretion of resin – resinosis – can be produced by injury or fungal infection, and also as a defence against many species of insects. Trees are more vulnerable in the peripheral regions of their species habitat during sudden changes of climate. In all these circumstances resin produces defensive compounds.

In the temperate zones, pines are some of the highest producers, but in the tropics the best resin-producers belong to the genus *Hymenaea* (Langenheim, 1999).

The Origin of Amber

The origin of amber may be traced back to plants that are members of the Gymnosperms and Angiosperms. Since 1835, various authors have proposed, on the basis of morphological and anatomical studies of plant inclusions in amber and in amber-bearing deposits, species of plants that could be responsible for the production of amber, in the genera *Pinus* (*Pinus succinifera*), *Pinites, Abies, Taxoxylum, Pityoxylon*, and *Picea*. Only two types of trees living today produce stable resins that could, in time, fossilize into amber: they are the Kauri pine (*Agathis australis*) of New Zealand and species of the legume *Hymenaea* in east Africa and south and central America (Ross, 1998).

Amber can be dated on the basis of the fossils in the associated sediments. Animal and plant inclusions in amber are priceless material for paleontologists. In fact, resin is probably the best medium of all for preservation of organisms that become entombed in the sticky material. However, if amber is re-worked (eroded from one sedimentary deposit and re-deposited elsewhere) then the amber could be much older that sediments suggest. The frequent presence of *Sequoioxylon gypsaceum* both in the amber-bearing formation and in romanite itself (Petrescu et al., 1989) suggests that this conifer could have been the source of the ancient resin that has fossilized into the present-day amber occurring in Oligocene deposits at Colti, Romania. Palaeobotanical and palynological researches on the amber-bearing formations of Colti indicate the age as Upper Rupelian-Early Chattian (Ghiurca and Vavra, 1990).

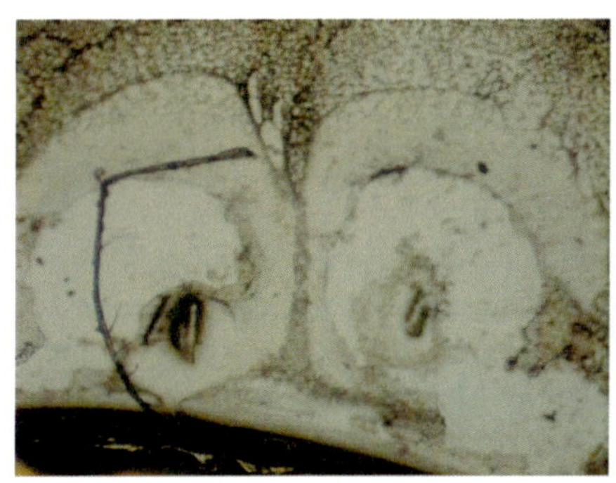

Sequoiapollenites in romanite (Neacsu, 2008)

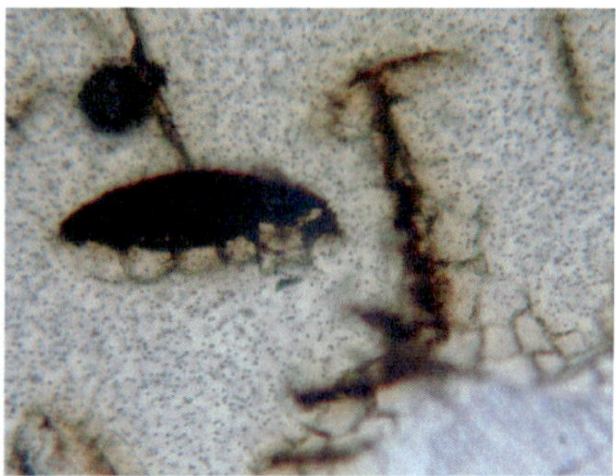

Insects Ord. Coleoptera and fluid inclusions in Baltic amber, Poland (Neacsu, 2008)

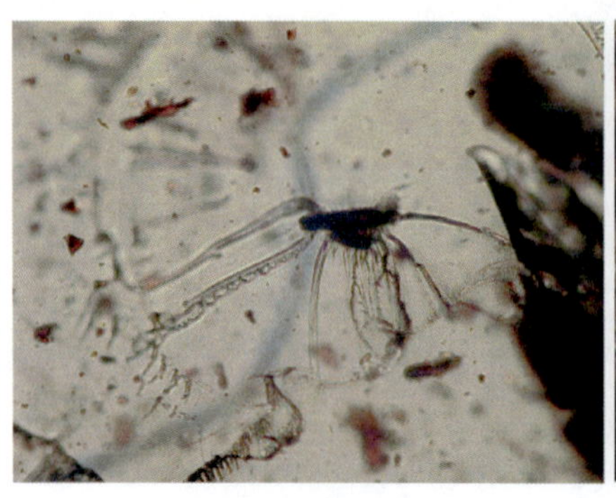

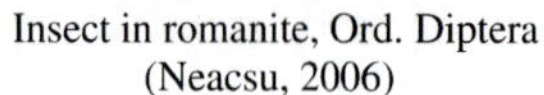

Insect in romanite, Ord. Diptera
(Neacsu, 2006)

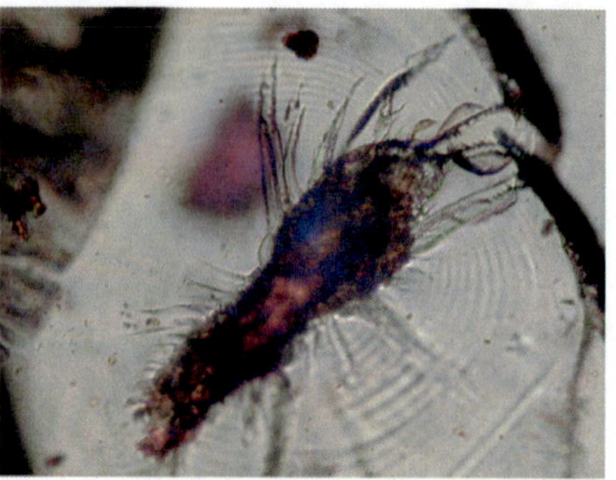

Insect in romanite, Ord. Coleoptera
(Neacsu, 2006)

Chemical Structure of Amber

Amber is composed of high-molecular weight volatile compounds, including aromatic hydrocarbons (cymenes), terpenoids and polymers, and their oxidation products. It cannot be described by a single chemical formula. Baltic amber is also known as succinite, due to the high content of succinic acid (up to 8%). This is not an original component of amber, but a degradation product of abietic acid (Rottlaender, 1970). Based on chemical studies, Ghough & Mills (1972) and Beck (1993) in Banerjee et al. (1999) state that it is not abietic acid or its derivatives, but bicyclic diterpen acid of the labdanum-type, which is the fundamental component of amber. This means that *Pinus* resin, characterized by a high abietane content, does not seem to polymerize or survive on geological timescales. Resins with a high labdatriene content (*fam. Cupressaceae*, some of *Taxodiaceae* and most *Agathis sp.*) start to polymerize as soon as they are yielded by the trees; when produced in sufficient quantity they survive for long periods of time.

Modern Techniques Used to Study the Geological Origin of Amber

Amber has become a most satisfying substance for interdisciplinary study. In Romania, the main aim is to clarify the geological origin of romanite and to establish whether it represents an amber species or it is only one of the many European amber varieties (Teodor et al., in press). The role of optical microscopy in the study of fossil resins is of primary importance. Apart from efficiency and ease of use, it offers a precision comparable only with chemical analyses; at the same time it allows detailed mineralogical investigations of the resins and of the host-rock. Diagenetic processes played a significant role in the creation of romanite. One of the most important consequences is the appearance of a tendency towards internal organization (Neacsu, 2006; Neacsu and Dumitras, 2008), proved by its weak anisotropy. Another consequence is shown by remineralizations in romanite (substitution of organic matter by anhydrite and feldspar). Two different romanite types exist: an older one, with distinct fissures, impregnated with organic material, and a recent one, lighter, which is adjacent to the former one and penetrates its fissures.

The X-ray diffraction data, confirmed by microscope studies, seem to indicate that romanite has some internal organizing tendency (Neacsu, 2006, 2008). Some crystalline components may give rise to diffraction patterns and could have a genetic significance.

Infrared spectrometry has revealed certain similarities between the absorption levels of beams of light in both Baltic amber and resin from modern-day *Cedrus atlantica*. Some researchers (Langenheim, 1995 in Neacsu, 2003, unpublished) have pointed to the similarities in the chemical composition of succinite and the resin of kauri trees – *Agathis australis, Araucariaceae* family.

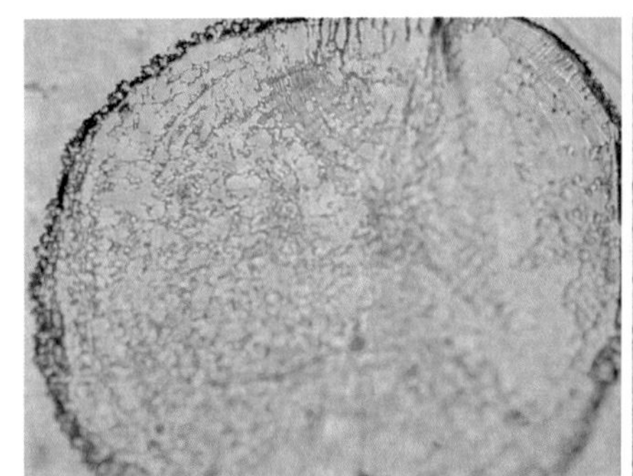

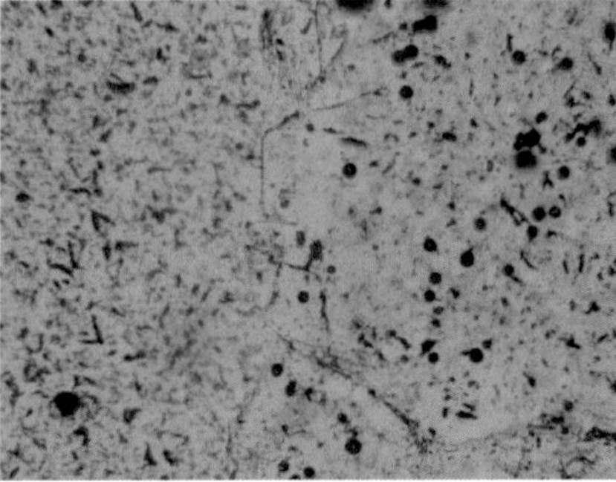

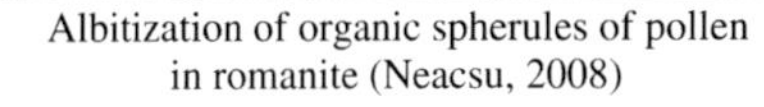

Albitization of organic spherules of pollen in romanite (Neacsu, 2008)

Contact of two romanite stages and glauconite (Neacsu, 2008)

The Exploitation of Amber

Succinite is by far the earliest and well known of all fossil resins. Resin secreted by coniferous trees at least 40 million years ago was carried southwards by rivers from the Scandinavia and the present-day Baltic Sea and laid down in Tertiary marine sediments known as *blue earth*. On the Sambian Peninsula (Russia), Baltic amber has been extracted from a deposit at a depth of 50 m almost continuously since the 17th century (Kosmowska-Ceranowicz, 2003). The amber-bearing Sambian Delta extends into Polish territory. Poland has reported a reserve of approximately 643,820 tons, with individual layers containing from 132 to 5,976.5 g/m^3. The deposits belonging to Poland lie at a depth of 120 m, which, for the time being, precludes their extraction (Kosmowska-Ceranowicz). Poland has become one of the leading producers of amber jewlery, processing some 200 tons annually from accumulations in Quaternary sediments of the Parczew Delta. The resources of the amber layer were estimated here at 6,911 tons.

In Romania, a beautiful amber was exploited from the 19th century, near Colti (Buzau County), on the Sibiciu Valley. The resin-bearing strata belong to the Oligocene of the East Carpathians flysch (Kosmowska-Ceranowicz, 1999). They are intercalated within the lower and medium parts of the lower Kliwa sandstone, having a thickness that is not constant (0.20–1.40 m). They consist of siliceous clay always containing thin intercalations of bituminous shales (2–5 cm) and of preanthracite coal (1–2 cm).

As a symbol of Romania, the brown–reddish amber was shown at the Universal Exposition in Paris (1867). Its systematic exploitation began in 1920, carried out by the engineer Dumitru Grigorescu. During 1923–1925 more than 300 kg of amber were extracted; by 1935 the exploitation ceased. After 1935 the amber was occasionally exploited until 1948, when the Ministry of Mines closed all such activities. Between 1981 and 1986 the Ministry of Mines and Oil restarted the amber exploitation, but the extracted material was lost or became part of private collections (Ciobanu and Dicu, 2005). Today amber from the Colti area is not for sale, the small quantities still found being used for scientific studies. People continue to look for amber along rivers after rain and usually in Spring time. Small amber nodules of various shape, ranging from a few milimeters up to 10 cm in size, may be found, rarely reaching 2–3 kg. The largest pieces are used for making jewels.

In the County Museum in the town of Buzau, one of the largest pieces of romanite in the world, weighting 3.450 kg, may be seen.

A large variety of crude and processed amber objects are displayed in a unique Amber Museum that was opened in 1980 at Colti.

Amber Museum, Colti, Romania

Amber pieces in the Colti Museum

Processed amber (jewels) in the Colti Museum

Since romanite is extremely rare and attractively coloured, it has become expensive on the European market.

References

Banerjee, A., Ghiurca, V., Langer, B. & Wilhelm, M. 1999. Determination of the provenance of two archaeological amber beads from Romania by FTIR-and solid-state-Carbon-13 NMR Spectroscopy. *Archäologisches Korrespondenzblatt*, 29, heft 4. Verlag des Römisch-Germanischen Zentralmuseums, Mainz

Ghiurca, V. & Vavra, N. 1990. Occurrence and chemical characterization of fossil resins from Colţi (District of Buzău, România). *N. Jb. Geol. Paläont. Mh.*, H.5, 283–294. Stuttgart

Ciobanu, D. & Dicu, A. 2005. Chihlimbarul, bijuterie şi elixir, Muzeul Judeţean Buzău

Ghough, L.J. & Mills, J.S. 1972. The composition of Succinite (Baltic amber). Nature, London

Kosmowska-Ceranowicz, B. 1999. Succinite and some other fossil resins in Poland and Europe, *Estudios del Museo de Ciencias Naturales de Alava*. 14, p. 73–117

Kosmowska-Ceranowicz, B. 2003. Amber from liquid resin to jewellery Ed. J. Popiolek, Bucureşti

Neacsu, A. 2006. *Remarks on the geological origine of rumanite. Acta Universitatis Szegediensis Acta Miner.-Petrogr.*, 5, 83, Szeged

Neacsu, A., Dumitras, D.G. 2008. Comparative physico-mineralogical study of romanite and Baltic amber; preliminary FT-IR and XRD data *Rom. J. of Mineral Deposits*, v. 83, p. 109–115, IGR Bucuresti 2008

Petrescu, I., Ghiurca, V. & Nica, V. 1989. Paleobotanical and palynological researches on the lower-oligocene amber and amber-bearing formation at Colti – Buzau. The Oligocene from the Transylvanian Basin, 183–198. Cluj-Napoca

Ross, A. 1998. Amber the natural time capsule, The Natural History Museum, London

Rottländer, R.C.A. 1970. On the formation of amber from Pinus resin. *Archaeometry*, 12

Shashoua, Y. 2002. Degradation and inhibitive conservation of Baltic amber in museum collections. Department of Conservation. The National Museum of Denmark

Teodor, E., Litescu, S.C., Neacsu, A., Truica, G. & Albu, C. 2008. Analytical methods to differentiate Romanian amber and Baltic amber for archaeological applications, *Central European Journal of Chemistry*, V. 7, no.3, p. 560–568, Springer September 2009, ISSN 1895–1066.

www.pan.pl

www.whyfiles.org/008amber/ambermain

Muddy Volcanoes

Cristian Marunteanu and Dumitru Ioane

The "Muddy Volcanoes", a natural wonder of the Vrancea zone, is a geological reservation located in the territory of Berca, Buzau county.

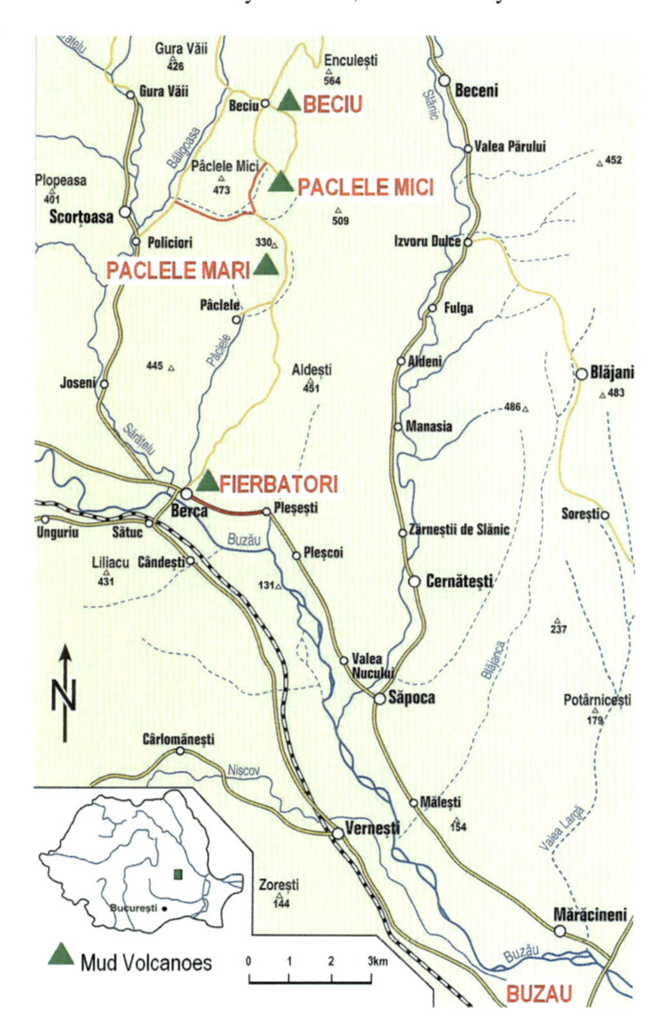

C. Marunteanu (✉)
University of Bucharest, Romania
e-mail: crimarunteanu@yahoo.com

N. Evelpidou et al. (eds.), *Natural Heritage from East to West*,
DOI 10.1007/978-3-642-01577-9_9, © Springer-Verlag Berlin Heidelberg 2010

The mud volcanoes from the Berca region, that are called in Romanian "pacle" (meaning something between "smoky" and "foggy"), are the most important in Romania and are considered as the most impressive in Europe. They are located in the Berca-Arbanasi zone, a well-known oil-bearing structure, within the outer part of the bending zone of the East Carpathians. Four mud volcano areas, which are situated along a NNE-SSW tectonic lineament, are presently active: Paclele Mari, Paclele Mici, Fierbatori and Beciu. The main ones, Paclele Mari and Paclele Mici, have been established as natural reservations since 1924 (Dicu and Bortas, 2005).

A "mud volcano" is usually associated with hydrocarbon gases, mostly methane, that are migrating upward from accumulations situated at different depths in geological structures. The gases, accompanied by almost continuous eruptions of muddy "lavas", migrate toward the surface through faults, fractures or fissures, the clayey matter being mixed with waters from underground aquifers. Cone-shaped forms with craters on top are usually built at the surface, modifying the local topography on a daily basis.

Mud volcanoes are quite widespread in Europe, being described in Spain, Italy, Greece, Crimea and Kerch peninsulas, and the Caucasus region (Taman peninsula, Georgia and Azerbaijan), as well as in Romania. These natural degassing structures represent a major source of methane in the atmosphere (Etiope et al., 2004).

Conical edifices at Paclele Mari

From a geological point of view, the area is located on the inner side of the Carpathian foredeep, consisting of Miocene molassic sediments. The four gas-emitting sites are situated along the axial trend of a 18 km long anticline fold, including a salt diapir in its deep core, affected by systems of longitudinal and transversal faults and bearing significant oil accumulations (Paraschiv, 1979, 1984).

Mud flows and drying fissures (Paclele Mari)

These mud volcanoes areas take the form of elliptical plateaus, covered by clay "lava" products differing in age; they are continuously enhanced by outpourings of intermingled water, clay and small fragments of other types of sedimentary rocks. The active "craters" change locations from time to time, due to modifications in the upper parts of the fracture systems, leaving traces of older eruptive activity and opening new vents.

The Paclele Mari area is a geomorphological structure about 2,000 m long, 1,000 m wide and 100 m high. The surface covered with clay products delivered by the mud volcanoes is larger than 1.5 km^2.

The smaller Paclele Mici area is about 1,300 m long, 1,000 km wide and 60 m high, the surface covered with "volcanic" products being larger than 0.5 km^2 (Sencu, 1985).

Paclele Mici site

Small mud vulcano with a gryphon on the flank (Paclele Mici)

Mud volcano crater (Paclele Mici)

In both areas the fluid matter is usually ejected from conical edifices of various heights, reaching a maximum of 10 m, displaying apical vents and circular pools filled with muddy water and gas bubbles.

Bubbling circular pool (Paclele Mari)

Circular pool filled with muddy water (Paclele Mari)

From time to time, local explosive activity may be observed. In the Beciu area, located north of Paclele Mici, an eruption with a muddy column 1 m high was observed for 24 hours in November 1976. During the 30 days of eruption, about 5,000 tons of mud were ejected and deposited in this area. On the occasion of the strong 1977 Vrancea earthquake (M: 7.2) the muddy eruption was reactivated for 6 hours (Baciu and Etiope, 2005).

The slope of the cones is variable, from almost horizontal in case of flat structures, to a maximum of 40°, depending on the mud viscosity. Numerous characteristic structures are represented by circular pools filled with muddy water, their diameter ranging between 3 and 10 m. The ejected waters are salty, bromide-rich, resulting from mixing processes between connate and meteoric water. A thin layer of crude oil may float from time to time on the surface of the pool (Baciu et al., 2007).

Bubbling mud volcano with dark oxidized petroleum products floating on the surface (Paclele Mari)

A large range of colours contributes to the beauty of the Vrancea mud volcano landscape, especially at sunrise and sunset. The mud displays greyish and brownish colours, depending whether is wet or dry, or considering the particular type of clay that formed the outpouring "magma". The brown or grey borders of the craters may sometimes be embellished by white crust of crystallized salt and yellow sulphur.

The activity of the mud volcanoes generally consists of a continuous gas bubbling, water overflow and small eruptions of mud from vents. Usually, if an open fire is brought close to them, a flame will be seen because the bubbles are mostly made of methane.

Mud flow (Paclele Mici)

Mud vulcano with a gryphon on a flank (Paclele Mari)

At the lower part of the volcanoes, the muddy flows dry up, resulting in ephemeral shapes resembling Mesozoic reptiles, that are rapidly modified by rain and wind. During rainy seasons, water flowing on the slopes cuts numerous ditches and ravines, creating in this way a particular landscape.

Erosion forms (Paclele Mici)

Erosion forms (Paclele Mari)

Fierbatori mud volcano area is situated in the southernmost part of the anticline structure, near Berca, the total area covered with muddy lava being about 0.09 km^2. The ejected products are more fluid in this area and that is why only flat cones with muddy water may be observed. Specific to this area is active degassing of methane from dry soil.

A small area of ca 0.01 km^2, with craters of low viscosity bubbling mud, is present in the Beciu area, located in the northern extremity of the anticline structure.

The total surface covered by recent and old mud volcanic products on the Berca-Arbanasi hydrocarbon-bearing structure is about 2.5 km^2, the total methane output being estimated of at least 1,200 tons per year (Etiope et al., 2004).

Emanations of natural methane gas without water and mud, generating everlasting fires, may also be observed in the region. In the vicinity of the Andreiasu village, 50 km north of Berca, there is a small plateau subject to intensive degassing. Flames of such everlasting fires may be as high as 1 m.

On the barren sites of the muddy plateaus there live plant species characteristic of a salty environment.

Plants in the salty environment

Two of these plants, *Nitraria schoberi* (gardurarita) and *Obione verrucifera*, are very rare. Paclele Mici is the only place in Europe where these plants grow and so, for their protection, this site has also been declared a botanical reservation.

References

Baciu, C., Caracausi, A., Etiope, G. & Italiano, F. 2007. Mud volcanoes and methane seeps in Romania: main features and gas flux, *Ann Geophysics*, vol. 50, no. 4, 501–512.

Baciu, C. & Etiope, G. 2005. Mud volcanoes and seismicity in Romania, in *Mud volcanoes, geodynamics and seismicity*, edited by G. Martinelli and B. Panahi, NATO series .Springer Verlag, Berlin, 77–87.

Dicu A. & Bortas V. 2005. Mud volcanoes in the Buzau area. Victor B Victor Publ. House, Bucharest, pp. 135 (in Romanian).

Etiope, G., Baciu, C., Caracausi, A., Italiano, F. & Cosma, C. 2004. Gas flux to the atmosphere from mud volcanoes in Eastern Romania, *Terra Nova*, 16, 179–184.

Paraschiv, D. 1979. Romanian oil and gas fields, *Inst. Geol. Geoph, Tech. Ec. St., Ec.*, 13, 5–382.

Paraschiv, D. 1984. On the natural degasification of the hydrocarbon-bearing deposits in Romania, *An. Inst. Geol. Geofiz.*, LXIV, 215–220.

Sencu, V. 1985. Mud volcanoes from Berca. Sport-Turism Publ. House, Bucharest, pp. 21 (in Romanian).

Post-volcanic Phenomena in the East Carpathians

Alexandru Szakács

Introduction

Due to its peculiar geographical position and geological evolution and structure, Romania hosts a wealth of natural heritage of great value. Among those of geological interest, a large variety of post-volcanic features are most important, being some of the very few examples of ongoing geological phenomena.

"Post-volcanic phenomena" is a collective name for processes developing within and around areas of past volcanic activity after volcanism has definitely ceased. Although their causal relationships to the volcanism are not straightforward, the spatial connection and temporal succession between them is obvious. They basically consist of release of thermal energy and volatile substances in and around the area of occurrence of volcanic rocks long after the end of volcanic activity. They are commonly regarded as representing the effects of cooling and degassing of deep-seated magma chambers that previously fed volcanic processes. Surface manifestations include thermal anomalies and related thermal springs, dry carbon dioxide emanations, mineral springs and their related deposits (carbonate and silica sinters).

The occurrence and frequency of such phenomena are controlled by a number of local geological and geographical features, such as tectonics, lithology, subsurface water levels and movements, and topography. The combination of these factors determines unique features of post-volcanic manifestations locally giving rise to a large spectrum of landscapes and biological habitats. Anthropic intervention – traditional or modern – may add further value to or, to the contrary, may degrade these local microenvironments.

A. Szakács (✉)
Sapientia University, Cluj-Napoca, Romania; Institute of Geodynamics, Romanian Academy, Bucharest, Romania
e-mail: szakacs@sapientia.ro

N. Evelpidou et al. (eds.), *Natural Heritage from East to West*,
DOI 10.1007/978-3-642-01577-9_10,

Location

Post-volcanic manifestations are present, identified, mapped and inventoried in all areas of Neogene volcanism in Romania (i.e. in the Apuseni Mts, Oas-Gutai Mts, the so called "subvolcanic zone in the East Carpathians, and along the whole Calimani-Gurghiu-Harghita range). However, their most frequent and diverse occurrence is related to the youngest segment of Neogene volcanic activity in the East Carpathians, namely the South Harghita Mts. and environs, where they form a well-developed "mofettic aureole". Within it, the surface manifestations of post-volcanic activity are extremely abundant and diverse, constituting one of the most valuable natural resources of the area shared by Harghita and Covasna counties.

Geological Background

Neogene volcanism developed along the boundary between the Transylvanian Basin to the west and the East Carpathian fold-and-thrust belt to the east in the time interval of ca. 10.5–0.04 Ma (Pécskay et al., 1995), resulting in the construction of the ca.160 km long Calimani-Gurghiu-Harghita volcanic chain (CGH). The outstanding feature of CGH, which distinguishes it within the framework of the broader Carpathian-Pannonian system, is represented by an obvious along-arc migration of volcanism in that time interval. CGH consists of a NNW-striking row of closely spaced volcanoes (most of them composite volcanoes) built up of eruptive products (lavas and volcaniclastics) of mostly andesitic composition (Szakács and Seghedi, 1995) whose age is progressively younger towards the southeast end of the chain. So, the youngest volcanism is in the South Harghita Mts, with the most recent eruptive event occurring at Ciomadul volcano ca. 35–42 Ka ago (Moriya et al., 1996). South Harghita is particular within CGH, in that it unlike the northern segments. It cuts across the Carpathian folded structures, hence having a different local basement consisting of Lower Cretaceous flysch deposits (Szakács et al., 1993).

Post-volcanic Phenomena and Features

Post-volcanic phenomena manifested at the surface include dry CO_2 emanations from mofette (fumaroles), CO_2-rich mineral water springs, bubbling pools and swamps, thermal springs and spring-related deposits. Subsurface features consist of mineral groundwater in phreatic or pressurized aquifers, as well as underground mineral deposits in places brought to the surface by erosion.

Area of natural diffuse CO_2 emanations with killed vegetation

Entrance to the Büdös cave, Turia

Unattended Bödös cave mofetta, Turia, with CO_2 gas level shown by sulphur deposition on walls, used for gas-bath cures

These phenomena occur only sporadically inside composite volcanoes, or more precisely within their craters; they are far more abundant on their peripheries. Currently most of the post-volcanic manifestations are located in areas where subsurface rocks are not volcanic at all, consisting of local pre-volcanic basement rocks outcropping around the volcanic edifices. In the case of the "mofettic aureole" of the South Harghita Mts., except for a couple of mofette and mineral springs at Sântimbru-Băi, a few mineral springs and one mofetta in the Cucu volcano crater and the "Torjai Büdös" (Puturosul) mofetta, they are found in sedimentary rocks of the Cretaceous Flysch or of the Pleistocene fill of intra-mountain basins.

Mofette emit low-temperature gas mostly consisting of CO_2. The emanations come to the surface from deep reservoirs using fractures and faults crosscutting basement rocks only where they can avoid being captured by subsurface aquifers. As well as CO_2, which is the main gas component (typically 97–98%), mofette may contains H_2S, N_2, CO, H_2, CH_4 and trace amounts of noble gases (He, Ar, Rn). Only a few mofette in the region are in a natural state, most of them occurring within and around the Puturosul Mt., which is a remnant of a Pleistocene dacite dome belonging to the Ciomadul volcanic structure. They can be recognized by patches of dead vegetation and by the rotten-egg smell of the emanating gas (due to its H_2S content) on the surface, or by sulphur deposits on cave walls and by the same bad smell within rocky outcrops.

The large majority of mofette are no longer in natural state, having been captured and modified by human intervention within artificial "mofettas" (the local name for wooden gas bathing pools covered by simple artisan structures).

Mofetta at Tusnad-Bai

There are thousands of CO_2-rich mineral water springs in the region, some of them captured and modified artificially for consumption, and many others in a natural state, especially along creeks running on fractures. Their chemical compositions vary widely according to the rock types traversed below the surface. Fe typically oxidizes at emergences and precipitates as iron hydroxide minerals around mineral springs giving the place a rusty color.

Mineral water spring in a valley bottom with red rust from iron hydroxide deposition, used for local people for drinking. Turia brook

Mineral spring in the Olt valley between Tusnad-Bai and Bixad

In places emergent mineral water collects in low-lying areas as bubbling natural pools or swamps over surfaces of various extent, up to several hectares (such as the Buffogo swamp near Balvanyos). Swampy areas may have a number of bubbling pools with mineral waters of different color according to local input of other components, such as mud or iron hydroxides giving the landscape a spectacular allure. Many of them have been artificially enlarged and adapted to be used as open-air pools for medical treatment, which has a secular tradition in the area.

Traditional mineral-water "foot-bath" pools near Balvanyos

Current development of a private mineral water pool system near Balvanyo

Mineral spring deposits are commonly carbonate sinters (also called calcareous tufa) originating due to the pressure drop at thermal spring outlets. They form small mounds of layered spongy carbonate, marking the sites of inactive fossil spring emergences. Silica sinters are rare overall in CGH; they are more frequent to the north, being related to higher-temperature thermal springs.

Thermal anomalies are associated with higher-than-normal heat flux in areas of the most recent volcanic activity, where a subsurface magma-chamber has not yet reached thermal equilibrium with the surrounding rocks. This is the case for the strongest heat-flux anomaly in Romania located at Tusnad-Bai, where "mesothermal" waters (23°C) feed an open-air swimming pool.

Exploitation of Post-volcanic Phenomena as a Natural Resource

Local people were quick to see the value of mineral water springs and pools, and dry gas emanations, as natural resources. The tradition in using these resources in a simple and artisan way for both current domestic and medicinal purposes goes back to the Middle Ages. More extensive exploitation of them arrived in the eighteenth century when popular spas and health resorts were founded around clusters of high-discharge mineral springs located in picturesque zones such as Tusnad, Malnas, Bodoc, Valcele and many others. Some developed in time to become internationally recognized tourist destinations and/or treatment-centers (e.g. Covasna, Tusnad-Bai). Also, carbonate mineral waters started to be bottled and commercialized, some of them (e.g. Borsec) acquiring world-fame. The prosperity of these spas and resorts fluctuated according to the changing availability of the natural source itself (e.g. deceasing discharge of springs) and to economic and political conditions. Some formerly famous spas, such as Valcele, went completely down hill and vanished, while others, especially those with acknowledged curative properties and proper facilities, such as Covasna, kept themselves busy and attracted increasing interest.

Despite these examples, the full potential of post-volcanic phenomena as natural resources is far from being fully exploited in the region. Nevertheless, as recent beneficial developments one may mention the creation, mainly by the efforts of private and NGO organisations, of a "borviz" (the local Hungarian name for mineral water) track along the Southern Ciuc basin and a small "borviz" museum at Tusnad village attracting an increasing number of tourists. There is a lot to be done in infrastructure development and modernization of traditional tourist and treatment centers and in attracting new resources to the tourism and medical treatment circuit in a nature-friendly and ecological manner.

Conservation Status

Viewed as part of a valuable natural and cultural heritage, the post-volcanic manifestations of the South Harghita "mofettic aureole" are at different, although generally low, levels of conservation. In the more important spa areas they are to some extent integrated in the local historic and cultural heritage. However, no specific protection is legalized or being managed, except for technical purposes such as the hydrogeological sanitary protection areas around medically certified mineral springs. Some mofette are private-owned and their maintenance is supervised for health security reasons. A few are under development. Others are untouched. There is no legal reinforcement of rules of conservation and environment-friendly exploitation of this peculiar type of natural resources.

Most mineral springs in their natural state along valleys are not protected at all, and so are subject to all kinds of natural and anthropic degradation processes. As yet there is not even a full inventory of these resources. A team of professors and students from the Babes-Bolyai University, Cluj-Napoca, started a few years ago to complete such an inventory on scientific basis, but the work has very low funding and advances slowly.

References

Moriya, I., Okuno, M., Nakamura, T., Ono, K., Szakács, A. & Seghedi, I. 1996. Radicarbon ages of charcoal fragments from the pumice flow deposit of the last eruption of Ciomadul volcano, Romania (in Japanese with English Abstract). *Summaries of Researches Using AMS at Nagoya University*, VII, p. 252–255

Pécskay, Z., Lexa, J., Szakács, A., Balogh, K., Seghedi, I., Konecny, V., Kovacs, M., Márton, E., Kaliciak, M., Széky-Fux, V., Póka, T., Gyarmati, P., Edelstein, O., Roşu, E. & Zec, B. 1995. Space and time evolution of Neogene-Quaternary volcanism in the Carpatho-Pannonian Region. *Acta. Vulcanologica*, 7(2), 15–28

Szakács, A., Seghedi, I. & Pécskay, Z. 1993. Pecularities of South Harghita Mts. as terminal segment of the Carpathian Neogene to Quaternary volcanic chain. *Rev. Roum. Geologie.*, 37, 21–36, Bucharest

Szakács, A. & Seghedi, I. 1995. The Calimani-Gurghiu-Harghita volcanic chain, East Carpathians, Romania: Volcanological features. *Acta Vulcanologica*, 7(2), 145–153

Ancient Gold Mining in Rosia Montana (Apuseni Mts, Romania)

Dumitru Ioane and Horea Bedelean

Historical Background

Rosia Montana, or Alburnus Maior as was called by the Romans, is a famous ancient gold mining area located in the Apuseni Mts, at the western boundary of Transylvania, Romania. Its richness in gold was known and valued by ancient exploitations of river sediments or shallow veins long before the Roman conquest of Dacia (Motiu, 2004). In fact, the gold of Dacia was the very reason for the two wars between the Romans and the Dacians; in 106 AD. Dacia became a Roman province located at the northern limit of the empire.

Ancient Gold Exploitation

During the Roman administration (106–273 AD) mining activities were intensively developed. Numerous relics of mining works that are still preserved create a specific geomorphological landscape and represent a witness in stone of now legendary times.

Gold mining in Rosia Montana was greatly developed by the Roman administration, a network of mining galleries that followed the mineralized veins being carried out on several levels using hammer, vinegar and fire, mostly by colonists from Iliria.

D. Ioane (✉)
University of Bucharest, Romania
e-mail: dumitru.ioane@g.unibuc.ro

N. Evelpidou et al. (eds.), *Natural Heritage from East to West*,
DOI 10.1007/978-3-642-01577-9_11, © Springer-Verlag Berlin Heidelberg 2010

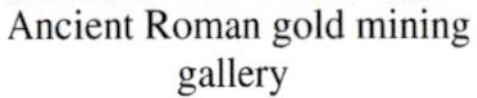

Ancient Roman gold mining gallery

Traces of Roman galleries at Rosia Montana

Ancient Roman gold mining gallery

Copies of ancient gold bars with Roman stamps are preserved in the Museum of Gold, established in the town of Brad, an important centre of gold exploitation located south of Rosia Montana.

Gold bars bearing Roman stamps Museum of Gold - Brad, Romania

Archaeological Finds

Archaeological finds, such as Roman mining galleries, lamps and parts of the drainage system, as well as funerary stones and votive altars are evidence for a vivid economic and cultural life in this remote region of the Roman Empire (Damian, 2003).

Roman archaeological relics in the Rosia Montana museum

Roman archaeological relics in Rosia Montana museum

Valuable samples of original Latin writing were discovered in the period 1786–1855, in the ancient mining works. Of the 50 written wax tablets found, 25 are still preserved in European museums.

Pieces of ancient andesite sarcophagus or Roman pottery may be still found when hiking in the hills near the old mines.

Andesite cover of an ancient sarcophagus

Roman pottery revealed by rain water in a road at Rosia Montana

Endangered Natural and Cultural Heritage at Rosia Montana

Weathering of the rocks and erosion processes over many centuries revealed numerous Roman mining galleries in the upper part of Cetate Hill, creating a picturesque and romantic environment. The resulted landscape looked very much like an old deserted fortress ("Cetate" meaning "Fortress" in Romanian) (Santimbrean and Bedelean, 2002).

Since 1970 modern mining activities have, to a great extent, destroyed this fascinating site, a large quarry having been excavated for gold exploitation. Despite this, after more than three decades of intense anthropogenic "erosion" using dynamite, there are still natural beauties and structures of great archaeological interest.

Violent destruction of the Rosia Montana "fortress"

Quarrying as a major enemy of natural & cultural heritage

In the neighborhood of Rosia Montana, several interesting geological sites are to be found, that have been established as "natural monuments"; among them are Piatra Despicata, Piatra Corbului, Detunata Goala and Detunata Flocoasa.

However the unique natural and archaeological site at Rosia Montana, still preserving the local geomorphology, is presently endangered by a recently proposed mining project that assumes quite rich gold mineralization at different depths. The ongoing debate about the fate of Rosia Montana is both national and international, and the official final decision has not yet been made.

Two main options are presently envisaged for Rosia Montana (Ioane et al., 2007):

- gold exploitation in a large and deepening quarry, that may bring jobs for some local inhabitants and revenues for the Romanian state for a limited interval of time, but will also damage the environment in both predictable or/and unknown ways;
- preservation of the natural, ancient mining and archaeological landscape and its promotion for active tourism, involving local inhabitants for a much longer period of time.

Acknowledgements Lucian Muntean is thanked by the authors for kindly offering some of the illustrations used in this paper.

References

Damian, P. (Editor). 2003. Alburnus Maior I. *ARTA GRAFICA Publishing House*, Bucuresti.

Ioane, D., Marunteanu, C., Bedelean, H., Oaie, G. & Santimbrean, A. 2007. Next future for Rosia Montana, Romania: geotourism or gold mining? *Integration of the geomorphological environment and cultural heritage for tourism promotion and hazard prevention, Italo-Maltese Workshop Abstracts Volume*, Malta.

Motiu, I. 2004. Dacia Provincia Augusti. *CORINT Publishing House*, Bucuresti.

Santimbrean, A. & Bedelean, H. 2002. Rosia Montana – Alburnus Maior, Cetate de scaun a aurului romanesc. *ALTIP Publishing House*, Alba Iulia.

Rupestrian Settlements in the Alunis Area

Cristian Marunteanu and Dumitru Ioane

Introduction

The rupestrian settlements located in the area of Alunis, Buzau County, Romania, are generally considered to have begun as ancient sites of refuge and cult in the Buzau Mountains. They are situated 6 km north-east of Alunis village and spread over an area of about 3 km^2.

Geological Setting

The area of rupestrian settlements is situated in the East Carpathians Bend Zone, the geological formations being represented by sedimentary deposits of the Tarcau Nappe, a regionally developed structure that overthrusts the Marginal Folds Nappe (Sandulescu, 1984). The outcropping Paleogene sedimentary deposits have been affected by high intensity tectonic processes as a consequence of continental collision, subsequent to a final stage of subduction along the East Carpathians.

The folded and faulted structures are trending NE-SW in this area (Dumitrescu et al., 1968), as are the active crustal faults involved in the high seismicity of the Vrancea zone, well known worldwide for its high magnitude earthquakes.

C. Marunteanu (✉)
University of Bucharest, Bucharest, Romania
e-mail: crimarunteanu@yahoo.com

N. Evelpidou et al. (eds.), *Natural Heritage from East to West*,
DOI 10.1007/978-3-642-01577-9_12,

Fracture systems and traces of erosion processes in sandstone cliffs

The rupestrian works have been carried out in sandstone cliffs or isolated blocks, the stone cutting being done mostly in sectors where open fracture or fissure systems allowed active weathering and erosion processes.

Rupestrian Inhabitants in History

The rupestrian settlements are now represented by a church and more than twenty shelters of various shapes and sizes.

Since this area contains the most important occurrence of amber in Romania, the ancient rupestrian settlements might be related to the exploitation of beautiful and valuable products of nature. Amber pearls have been found in burial sites belonging to the Monteoru Bronze Age culture (Zaharia, 1995).

The rupestrian shelters have been mentioned since the Middle Age, and recent studies consider that one of them, Fundu Pesterii (Cave Bottom), was inhabited during the sixth to fourth centuries BC, while others were built later, during the third – sixth centuries AD. Several stone shelters are considered to be built after the thirteenth century AD, being inhabited until the beginning of the nineteenth century, mostly by monks (Nica, 2002).

Written evidences of their religious lives may be found in texts carved on rupestrian house walls in Slavonic, the language used in Christian Orthodox service until the second half of the nineteenth century.

In many cases, access to the rupestrian shelters was intended to be difficult, wooden ladders being usually used to climb the cliffs to reach the entrance. This kind of "bird nest" location shows the need of the inhabitants, who historically tried to find refuge or spiritual enlightenment, for solitude, isolation and protection.

Texts in Slavonic on the inner wall of a shelter

Rupestrian shelter on a steep sandstone cliff

Above the entrance of the stone shelter there are traces of a former wooden roof and a ditch for protection against rain and snow. Such weather is normal in a mountainous rainy region, characterized by cold winters.

During peaceful times, stone shelters near roads or streams were also used by monks and occasionally by shepherds.

The usual inhabitants, the monks, lived a very simple and austere life, using beds dug in weathered sandstone.

Rupestrian shelter located at the base of the cliffs

Austerity in a rupestrian monk's room

The shelter windows resembled those of a Medieval fortress, being narrow and penetrating thick walls built of natural stone. Their small size protected the people against predators, bears and wolves being widespread in most mountainous areas in Romania.

On rainy or frosty days, access to the rupestrian houses was difficult and dangerous, so carved stairs were very useful to their long-term inhabitants. The size of the steps may easily have led to legends that giants lived in this region.

The church of Alunis village is both a religious and natural monument, being founded by two shepherds, Vlad and Simion, in 1274. It was documented in 1351 and was used as a monastery until 1871 (Nica, 2002).

This church has its altar located within the sandstone cliff, while the rest of it is built of wood and painted blue on the exterior.

Window of a stone shelter room

The rupestrian church of Alunis village

Giant steps dug in stone

References

Dumitrescu, I., Sandulescu, M., Bandrabur, T. & Sandulescu, J. 1968. Harta geologica scara 1: 200.000, foaia Covasna. *Institutul Geologic*, Bucuresti.

Nica, D. 2002. Asezarile rupestre de la Alunis-Nucu, *Judetul Buzau.*

Sandulescu, M. 1984. Geotectonica Romaniei. *Editura Tehnica*, 335 p.

Zaharia, E. 1995. The Monteoru Culture. Treasures of the Bronze Age in Romania, *Ministerul Culturii*, Bucuresti.

Salt Karst in Manzalesti – Romania

Cristian Marunteanu and Dumitru Ioane

Salt Formation in Manzalesti

The salt in Manzalesti is localized in the Vrancea hilly region, within the East Carpathians Bend, north of the town of Buzau.

The Manzalesti zone, where the salt massif bearing the same name outcrops, is situated in the Magiresti–Perchiu Digitation, the innermost tectonic unit of the Subcarpathian Nappe (Săndulescu et al., 1980; Săndulescu, 1984).

This tectonic unit, bordered to the west by the frontal part of the Marginal Folds Nappe and to the east by the Carpathian Foredeep along the Casin – Bisoca Fault, includes, in normal stratigraphic succession, the following sedimentary deposits:

- the bituminous lithofacies of the Kliwa Sandstone (Oligocene–Early Miocene), outcropping north of the Manzalesti zone;
- the salt formation (Early Burdigalian), composed of massive argillaceous breccias, discontinuous gypsum strata and salt massifs;
- the Magiresti Formation (Burdigalian), represented by an alternation of reddish or green clays, marls and sandstones;
- the grey schlier formation (Late Burdigalian – Early Badenian), presenting great lithological variability (clays, marls, sands, sandstones, gypsum beds or lenses, tuffites and thin laminated dolomitic or calcareous shales) and containing two marker levels, the Perchiu Gypsum at the base and the Stufu Gypsum in the uppermost part of the formation;
- the Slanic Tuff (Early Badenian), constituted of white tuffaceous marls (with abundant microfauna and nanoflora), interbedded with dacitic tuffs and tuffites.

The Early and Middle Miocene deposits have a folded tectonic style.

The salt formation outcrops 2.5 km west of the village of Manzalesti and is represented by a beautiful salt massif that develops over 3 km toward the NNE. The

C. Marunteanu (✉)
University of Bucharest, Bucharest, Romania
e-mail: crimarunteanu@yahoo.com

N. Evelpidou et al. (eds.), *Natural Heritage from East to West*,
DOI 10.1007/978-3-642-01577-9_13,

massif is surrounded by a grey schlier formation and represents a false salt diapir, a result of an injective folding (Stille's model – Săndulescu, 1984). In this tectonic process, the salt rises toward the surface through a reverse fault.

Karst Developed on Salt

Salt occurs frequently in Romania in the Miocene formations. The salt massifs sometimes outcrop, but in most situations they are covered by sedimentary rocks. Typical dissolution forms, such as lapies (karren), dolines, natural bridges or even blind valleys, develop where the salt massifs outcrop.

Outcropping salt in the Manzalesti area, Buzau County

Lapies in the salt massif

Salty spring and lapies

Karst collapse in salt

When the salt massif is the bedrock of another sedimentary formation, collapse or suffusion phenomena can occur, affecting the relief surface by underground salt dissolution, generating dolines or karstic sinkholes and valleys. Due to the rapidity of the dissolution processes, these relief forms are frequently ephemeral.

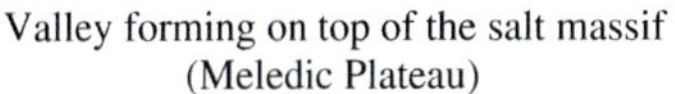

Valley forming on top of the salt massif (Meledic Plateau)

Small canyon in the salt massif

In Romania there are several salt caves, a few of them being located in the Manzalesti area of Buzau County. The biggest salt cave at Manzalesti is 3,234 m long and 42 m high. It is thought to be the second longest salt cave in the world. This cave is crossed by salty streams and offers, together with the particular scientific interest, a special range of extremely colorful formations.

Entrance to a salt cave

Sinkhole and cave in the salt massif

Meledic Plateau

The Meledic Plateau is a geological, spelaeological, botanical and zoological reservation. The area of scientific and tourist interest covers 67.5 ha in the territory of Manzalesti village, on top of the salt massif.

The caves resulting from salt karst phenomena occur in the salt massif that is overlapped by the Meledic Plateau. In places the salt massif is covered by thick clayey soil that hosts fresh water lakes.

In areas with salt situated at shallow depths, the vegetation is composed of halophyte plants. The warmer sites are generally occupied by sub-thermophile or even thermophile plants. The reduced forest cover and the NE-SW orientation of the relief features create a favorable habitat for Mediterranean fauna (scorpions, turtles).

Fresh water lake on the salt soil cover (Meledic Plateau)

Piatra Alba

Piatra Alba (White Stone) is a natural monument of geological interest, covering an area of about 0.025 ha. It is also situated in the territory of Manzalesti village and consists of volcanic tuff (Slanic Tuff).

This natural monument, located at the confluence of the Slanic and Jgheab streams, represents a witness of hydraulic and aeolian erosion.

Piatra Alba: witness of volcanic tuff erosion

Piatra Alba – the "White Church" on the Slanic river

References

Săndulescu, M., Micu, M. & Popescu, B. 1980. La structure et la paléogeographie des formations miocènes des Subcarpathes Moldaves. *Ass. Geol. Carp. Balk., Congr. IX, Tekt.*, p. 184–197, Kiev

Săndulescu, M. 1984. Geotectonics of Romania (in Romanian). *Editura.Tehnica*, p. 336, Bucharest

Montesinho and the Mountains of Northern Portugal

Tomás de Figueiredo

Portugal has a singular geographic position in the European context (Ribeiro et al., 1987). Set on the most western coast of the Iberian Peninsula, Portugal faces the Atlantic in an area where it is already characterized by the cold and rough waters so well known to northwestern Europeans. This coast looks due west, as can be seen from the continental map outline; and it certainly contributed to Portugal's leading role in the fifteen to sixteenth centuries and later. As a maritime power Portugal provided Europe with the wonders from across the seas, and as a cultural melting pot Portugal, at that time, reshaped the European view of the world. The vision of a maritime destiny drove the policies and common thinking of Portugal for a long time. Portugal shares the Atlantic with other western European countries; and nowadays also shares their strategic agenda, in which marine environmental threats and sustainable resource use rank highly (Ferreira, 2005b).

Portugal, a singular geographic position: facing the Atlantic, rooted in the Mediterranean world

T. de Figueiredo (✉)
Instituto Politécnico de Bragança, Escola Superior Agrária, Mountain Research Center – CIMO, Campus Sta Apolónia, 5301-855, Bragança, Portugal
e-mail: tomasfig@ipb.pt

N. Evelpidou et al. (eds.), *Natural Heritage from East to West*,
DOI 10.1007/978-3-642-01577-9_14,

In spite of its peripheral position, Portugal is part of the Mediterranean world (Ribeiro, 1986). Mediterranean climatic features are found in almost all the Portuguese continental territory. The exception is the northwestern tract, from the coast to an impressive mountain range inland, whose crests are the wettest spots in all Europe and where Mediterranean character is drowned out by the western oceanic influence (Azevedo et al., 1998). Climate broadly determines vegetation distribution and thus Portugal hosts the flora also found around the Mediterranean basin, although with specificities and endemic features that give it a characteristic richness and variety (Guerreiro, 1991). Crops, crop systems and cultivation techniques also came from the East (Caldas, 1998). The cultural matrix of Portugal has been influenced from far. Greek, Phoenician and Carthaginian settlers, Roman, Germanic and Arab conquerors: all sought the products of this land; all left their intangible imprint and material legacy; all added a layer, thicker or shallower, to the firm cultural ground Portuguese step on. And all but the German came from the Mediterranean basin (Mattoso, coord. 1992).

The olive tree, symbol of the Mediterranean: on a grove at Valbom dos Figos, near Mirandela, NE Portugal (photo by A. Guerra)

In Portugal, Atlantic and Mediterranean, the sea is an ever-present geographic feature (Medeiros, 2005). The coastal sands and cliffs, however, form a narrow strip, and sea winds are hardly felt inland, where continental effects originating in the core of the Iberian Peninsula progressively replace those of the ocean, and seasonal climatic contrasts become wider (Daveau, 1985). Orography is the main factor explaining climatic gradients from the coast inwards, as altitude increases and mountain ranges block the eastward trajectory of the humid air (Ferreira, 2005a).

However, Portugal is not globally a highland country, as 70% of the territory is below 400 m, average elevation being 240 m (Medeiros, 1987). But the North (average 370 m) and the South (160 m) clearly differ in elevation and any orographic effect on climate mainly concerns the area north of the Tagus River, where 75% of the land is above 200 m. The maximum elevation (1,993 m) is in the top of Serra da Estrela, in the Central Cordillera, whereas the south the highest Serra barely exceeds 1,000 m (S. Mamede in eastern Alentejo, 1,027 m).

The Northern Portugal Highlands are thus the most important Portuguese mountain areas. They all belong to the same geo-structural unit: the Hesperic Massif. The other two units are the Mesocenozoic rims (western and southern belts with, dominantly, Secondary limestone and sandstones) and the Tagus basin (actually the

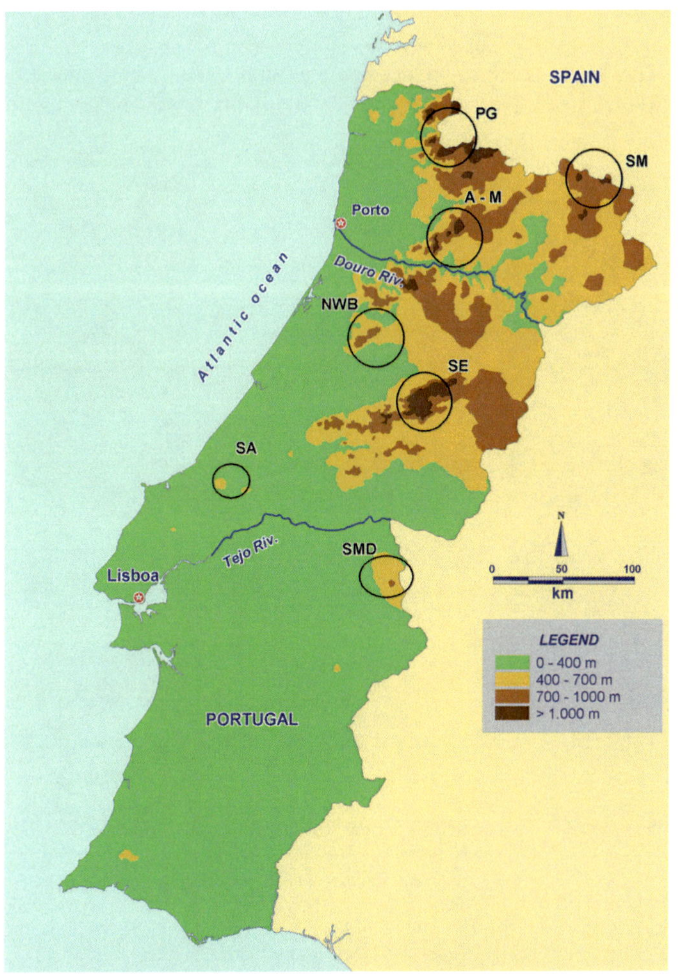

Hypsometry and the highlands of Portuguese continental territory (SM – Serra de Montesinho; AM – Alvão-Marão; SE – Serra da Estrela; NWB – Northwest Beira Highlands; PG – Peneda-Gerês; SA – Serra d'Aire; SMD – Serra de S. Mamede) (adapted from Ribeiro 1986; map by N. Evelpidou)

ancient basement depressed and covered with thick Tertiary or younger sediments, mostly loose). The Hesperic or Ancient, Paleozoic basement outcrops in most of the Portuguese continental territory. However, north of the Tagus, it is part of the Iberian Meseta northern block, a large plateau where average altitude is around 700m, in contrast with the southern block that does not exceed 400 m in elevation (Medeiros, 1987). Hypsometry is the outcome of this structure and of its very long history, which dates back to Paleozoic times and includes Hercinian (Variscan) orogenic activity and later lithospheric block rearrangements (Medeiros, 1987). Sedimentary rocks barely lasted so as to contribute significantly to the lithology of the Northern Portugal Highlands, which are composed of metamorphic (schists) and magmatic rocks (granites). Prevailing morphology and lithology do not favour pedogenesis, and so the soils are, for the most, incipiently developed. The shallow, stony and acid soils of these areas have low fertility and are generally only suitable for forestry or other less resource-demanding land uses (CNROA 1983).

Cold, windy and remote, having few resources that would attract and invite human settlement, the Northern Portugal Highlands retain the remnants of those

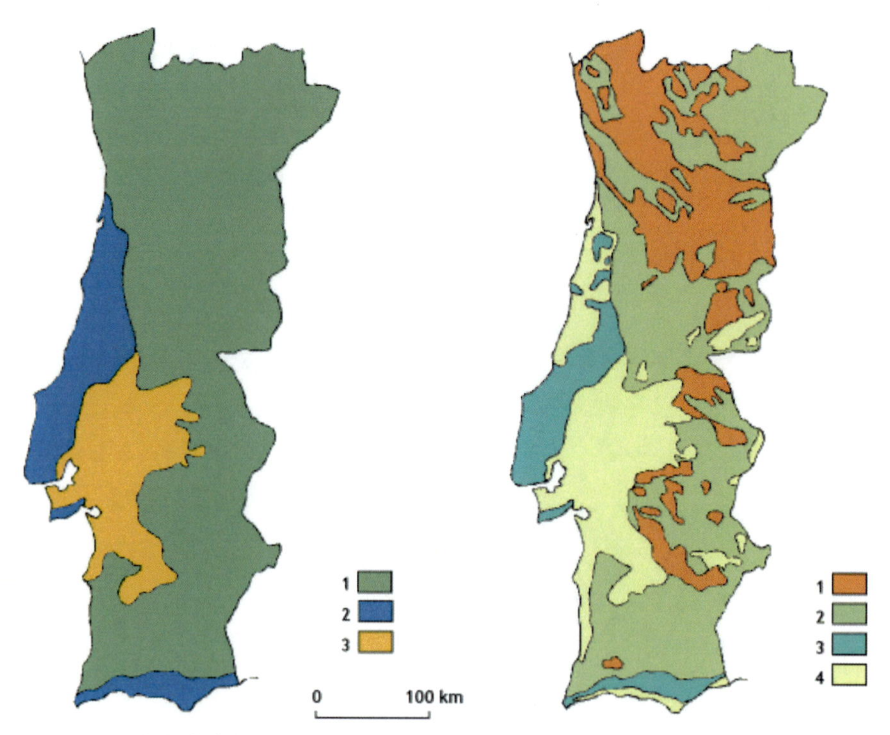

Geological sketch of continental Portugal (very simplified): Left – Morpho-structural units (1 – Hesperic Massif; 2 – Mesocenozoic Rims; 3 – Tagus-Sado Cenozoic Basin); Right – Main lithologies (1 – Variscan acid igneous rocks, as granites; 2 – Paleozoic metasedimentary rocks, as schists; 3 – Mesozoic sedimentary deposits, mainly hard, as limestones, marls, sandstones and clays; 4 – Cenozoic sedimentary deposits, mainly loose, as sands, clays, sandstones, marls and limestones) (adapted from Ferreira 2005)

ancient people to whom the area, with its valleys containing thick barriers of tall bushes, was a refuge from a menacing world. The Castros and Citânias archeological sites give us a picture of life before the Romans, who took so long to make these brave and rude highlanders their subjects (Mattoso, coord. 1992). The mountains are still a home for the few prepared to live under harsh conditions, a territory for shepherds and moving herds till recent times, with relicts of communal land use and wide forest areas, not always sharing peaceful co-existence (Ribeiro 1991, Caldas 1998). This is a changing land, less and less populated, more and more sought after by visitors seeking recreation, adventure, nature, tradition and culture. It is also a land increasingly in demand for energy production: hydropower, wind energy and biomass. New challenges and conflicting interests are changing this land, while seeking opportunities to test sustainability and related concepts applied in practice.

Montesinho: shrubland near Aveleda, NE Bragança, with Spain in the far horizon (photo by T. de Figueiredo)

Montesinho is the most northeastern of the Portuguese mountains and the fourth highest (1,487 m). It gives its name to a Natural Park, one of the largest in Portugal. The Natural Park has a remarkable geology and relevant natural values, and includes singular examples of a balanced and long-lasting relationship between the human communities settled here, the land they use, and the landscape they helped create. This was the study area selected by the Portuguese team in EDUNatHer, a Leonardo da Vinci Pilot Project. The team is, in fact, part of the team responsible for the Management Plan of Montesinho Natural Park (IPB/ICN 2006); this is a document recently produced by the Polytechnic Institute of Bragança (IPB), under contract to the Institute for Nature Conservation (ICN). Furthermore, Montesinho is, or has been, a study area for a large number of IPB researchers during the last 20 years or more. Scientific expertise, experience in converting it to decision support documents, and the involvement in EDUNatHer, all together explain the importance given to Montesinho in this context.

Zêzere River in Serra da Estrela: a Tagus tributary running in a U-shaped old glacier valley (photo by C. Aguiar)

Montalegre area in Peneda-Gerês National Park (photo by J. Vicente)

Besides Montesinho, the chapter describes each of the major mountain ranges north of the Tagus River (excluding those on the Mesocenozoic western rim with a maximum of 679 m elevation, in Serra d'Aire). Gerês is the most northwestern and it is the most humid of all, with summit elevation at 1,545 m. It is in the oldest protected area of Portugal, the National Park of Peneda-Gerês. Although actually distinct, Alvão and Marão are two neighboring peaks (1,339 m and 1,415 m, respectively) with similarities that justify being treated together. If Montesinho, Alvão-Marão and Gerês are, strictly speaking, northern highlands, the others considered here have the same label in a wider sense. Serra da Estrela, or simply Estrela, is the highest peak in Portugal, and gives its name to the largest Portuguese Natural Park. The Northwest Beira (an old province name, with no current administrative meaning but still commonly used) comprises a series of summits with lower altitude than those just mentioned (Caramulo, the highest, 1,200 m); nevertheless this is a typical highland area.

Photos By

A. Guerra, alzira.guerra@gmail.com

References

Azevedo, A. L., Gonçalves, D. A. & Machado, R. M. A. 1998. *Enclaves de clima Cfs no Alto Portugal: a difusa transição entre a Ibéria Húmida e a Ibéria Seca* (Série estudos N 39). Bragança: Instituto Politécnico de Bragança.

Caldas, E. de C. 1998. *A Agricultura na História de Portugal*. Lisbon: Empresa de Publicações Nacional.

CNROA. 1983. *Carta de Capacidade de Uso do Solo de Portugal* (7th ed.). Lisbon: Centro Nacional de Reconhecimento de Ordenamento Agrário (Soil Mapping Office), Ministry of Agriculture.

Daveau, S. 1985. *Mapas Climáticos de Portugal. Nevoeiro e Nebulosidade, Contrastes Térmicos*. Lisbon: Centro de Estudos Geográficos (Centre for Geographical Studies), University of Lisbon.

Ferreira, A. de B. 2005. Formação do Relevo e Dinâmica Geomorfológica. In A. de B. Ferreira (coord) *Geografia de Portugal: I – O Ambiente Físico* (dir. C. A. Medeiros): 53–255. Mem Martins: Círculo de Leitores.

Ferreira, D. de B. 2005a. O Ambiente limático. In A. de B. Ferreira (coord) *Geografia de Portugal: I – O Ambiente Físico* (dir. C. A. Medeiros): 305–385. Mem Martins: Círculo de Leitores.

Ferreira, D. de B. 2005b. O Espaço Atlântico Oriental. In A. de B. Ferreira (coord) *Geografia de Portugal: I – O Ambiente Físico* (dir. C. A. Medeiros): 257–303. Mem Martins: Círculo de Leitores.

Guerreiro, M. G. 1991. *O Mundo Mediterrâneo: sua diversidade e seu futuro*. Loulé: University of Algarve.

IPB/ICN 2006. *Plano de Ordenamento do Parque Natural de Montesinho: I – Relatório de Caracterização*. Bragança: Instituto Politécnico de Bragança.

Mattoso, J. (coord.) 1992. *História de Portugal: I – Antes de Portugal* (dir. J. Mattoso). Mem Martins: Círculo de Leitores.

Medeiros, C. A. 1987. *Introdução à Geografia de Portugal*. Lisbon: Editorial Estampa.

Medeiros, C. A. 2005. O território e o seu conhecimento geográfico. In A. de B. Ferreira (coord) *Geografia de Portugal: I – O Ambiente Físico* (dir. C. A. Medeiros): 18–45. Mem Martins: Círculo de Leitores.

Ribeiro, O. 1986. *Portugal, o Mediterrâneo e o Atlântico* (4th ed.). Lisbon, Livraria Sá da Costa Editora.

Ribeiro, O. 1991. Montanhas pastoris de Portugal: Tentativa de representação cartográfica. In *Opúsculos Geográficos: IV – O Mundo Rural*: 257–272. Lisbon, Calouste Gulbenkian Foundation.

Ribeiro, O., Lautensach, H. & Daveau, S. 1987. *Geografia de Portugal: I – A Posição geográfica e o Território*. Lisbon: Edições João Sá da Costa.

Montesinho Natural Park: General Description and Natural Values

J. Castro, Tomás de Figueiredo, Felícia Fonseca, João Paulo Castro, Sílvia Nobre, and Luís Carlos Pires

The Physical Environment

The Montesinho Natural Park (PNM, *Parque Natural de Montesinho*) is a protected area located in the municipalities of Vinhais and Bragança, in the administrative NUT Alto Trás-os-Montes (PT118), the mountainous region of northeast Portugal. It was created in 1979 and consists of 748 km^2 of natural wooded landscape and traditional mountain agricultural landscape, with highly variable gradients. PNM lies in the vast northeast Trás-os-Montes plateau, with average altitude around 750–900 m, which is part of the Iberian Meseta northern block (Medeiros, 1987; Ribeiro et al., 1987). However, in PNM elevation ranges more than 1,000 m, from the lowest point in the River Mente (436 m), its western border, to the top of Montesinho, at 1,487 m. The main altitudinal belts correspond also to the main landforms found in the area.

Cytisus spp and *Erica* spp vegetation in a mosaic between França and Aveleda, with *Serra de Montesinho* on the horizon (photo by T. de Figueiredo)

Most of the territory is in two areas, both below 1,000 m. The largest one is between 700 and 1,000 m, where most of the plateau lies. It is well defined in

J. Castro (✉)
Instituto Politécnico de Bragança, Apartado 1172, 5301-855 Bragança, Portugal
e-mail: mzecast@ipb.pt

N. Evelpidou et al. (eds.), *Natural Heritage from East to West*,
DOI 10.1007/978-3-642-01577-9_15,

the east, especially in Alta Lombada, a surface around 900 m elevation, whose "dark dorsal, poorly vegetated, with its regular outline, closes the horizon Bragança eastwards." (Taborda, 1987). In the west, the plateau is made of small platforms; these are transitional tracts between deep valleys and the neighboring highlands (Agroconsultores e Coba, 1991).

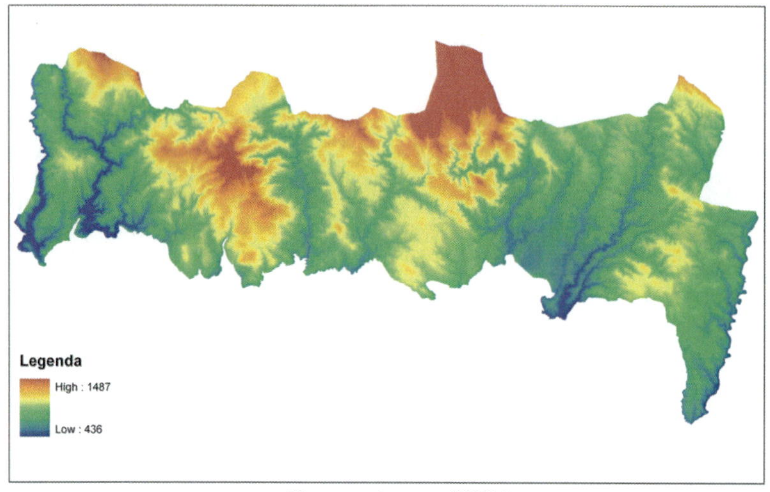

Hypsometric map of PNM

Most of the rivers in the area run below 700 m, and their valleys are the dominant landform in this area. Deep, with narrow bottom and steep slopes, i.e. V-shaped,

Tributaries of Sabor river in PNM: left) Onor crossing Rio de Onor, a village of communal traditions; right) Maçãs in Quintanilha, the eastern international border (photos by A. Suarez and A. Carvalho, respectively)

they strongly contrast with the planar or rounded crests in the interfluves. Also below 700 m, Baixa Lombada is a large, fertile, low-lying area north of Bragança (Gonçalves, 1980). It is a tectonic depression that contrasts with the high plateau of Alta Lombada, its eastern neighbor (Pereira et al., 2003).

In the mountains, above 1,000 m elevation, human occupation is almost nonexistent and hence cultivated land almost disappears, giving way to highland pastures (Agroconsultores e Coba, 1991). This contour bounds the mountain domains and surrounds the three highest peaks of the area, the southern edges of the Spanish Galician-Leonese mountains (Pereira et al., 2003): Corôa (1,272 m), Nogueira (1,318 m, in fact outside but close to the southern border of PNM) and Montesinho (1,487 m). The summits of these three mountains follow a morphological pattern found all over Trás-os-Montes: leveled hilltops, small areas of land corresponding to old erosion surfaces (Taborda, 1987).

Since it is an area of ancient rocks, some of the oldest in Portugal (see article by Meireles in this chapter), PNM reflects in its relief a remarkable morphogenetic activity, mainly due to the erosive action of running water, reshaping the surface and redistributing material in the landscape (Agroconsultores e Coba, 1991; IPB/ICN, 2006). Hypsometry and hydrography have thus been coupled in forming the land, on which plant communities have installed themselves, soils have deepened, humans have settled and which is finally opening to tourism and the interest of entrepreneurs.

Streams drain the whole area to the Douro River, more than 100 km to the south. The Douro is the Portuguese river with the largest drainage basin (though 4/5 of it is in Spanish territory), and it ranks second in total length (after the Tagus). The Douro drainage basin accounts for 1/3 of the Portuguese surface water resources and half of the energy generated in hydropower plants (Ferreira, 2005). Adding to its economic importance, the Douro drainage basin also hosts important terrestrial and aquatic wildlife habitats, including those of PNM.

Headwaters of two major tributaries of the Douro are in Montesinho, Tua and Sabor, the catchment divides being in fact across the border, in Spain. The Rabaçal and the Tuela, joining to form the Tua some 50 km to the south, are the largest rivers on the western side of PNM. The drainage basin of the Baceiro, a local tributary of Tuela, provides very significant examples of balanced land use in humanized landscapes (Gonçalves, 1980; IPB/ICN, 2006). The Sabor basin covers the eastern half of PNM and comprises the Sabor River itself (the largest in the area, with the largest drainage basin of Douro right bank tributaries), and the Maçãs, which drains the driest tracts of the Sabor basin.

In the area, as a whole, mean annual river discharge is 1/3 of precipitation; however, catchments have higher water yield in the west than in the east (Gonçalves, 1985). Following a common trend, explained by precipitation and vegetation cover decrease, inter-annual and seasonal variability of river discharge tends to increase from west to east (Henriques, 1985). Drainage density is high in the area, where stream networks are mainly controlled by geological fracture distribution (Pereira et al., 2003). The steep slopes of catchment surface and riverbeds, prevalent in this mountain domain, affect runoff and discharge, determining the habitat for aquatic species and stream ecology (see below). The effect of geological basement on catchment characteristics is clear when comparing drainage texture over granite and schist areas, the two main lithologies prevailing in PNM (Figueiredo and Fonseca, 1997).

Climate in Montesinho is Mediterranean, although other influences, derived from the geographic position and the relief of this area, affect the general pattern (Gonçalves, 1985). Continental effects come from the inner Iberian Peninsula, as Atlantic influence is hampered by impressive mountain ranges (more than 1,500 m elevation in the west and more than 2,000 m in the north, in Sanabria, Spain, close to the border). They bring an increase in seasonal contrast (annual range is 15–20°C), in dryness, and in inter-annual variability (Gonçalves, 1985). Continental effects are relevant in the eastern tract of PNM but they are drowned out to the centre and west and where the mountain effects take their place. Altitude is in fact the factor that best explains spatial variations in temperature (most of the area is below 12°C) and precipitation (mostly over 800 mm) (Gonçalves, 1985; Figueiredo, 1990).

Regional climatology labels Montesinho as "Terra Fria" (cold land), as opposed to "Terra Quente" (hot land), typical of the southern depressions and valleys less than 400 m elevation (Gonçalves, 1991b). In "Terra Fria" (annual temperature, $T < 12.5°C$), there is almost no transition between a short and hot summer and a long lasting winter, with the largest frosty season in Portugal (from mid-October to mid-May) (Agroconsultores e Coba, 1991; Gonçalves, 1991a; Ribeiro, 1996). Expressively, "Terra Fria" in a regional saying is "Nove meses de Inverno e três de Inferno" (freely translated to "winter for nine months, hell for the other three"). Most of the area of PNM lies on a plateau landscape ("Terra Fria de

Snow cover in the top of Montesinho (photo by A. Suarez)

Planalto"), where annual precipitation is over 800 mm west of Bragança, although it is lower to the east. The mountain ("Montanha") and high mountain ("Alta Montanha") landscapes are the typical of these altitudinal domains (T 9–10°C and T < 9°C, respectively), where total yearly rainfall is 1,200 mm or more, rising above 1,400 mm in the highest crests, partly falling as snow in winter. Small patches of transitional conditions ("Terra de Transição", T > 12.5°C) are found in the valleys close to the southern boundary of PNM.

The history of a wide variety of outcropping rocks (e.g. schist, granite, mafic and ultramafic rocks, unconsolidated sedimentary materials) makes PNM such a striking territory (see article by Meireles in this chapter). From them, soils have formed bearing vegetation; and both have developed in a wide variety of topographic and climatic conditions.

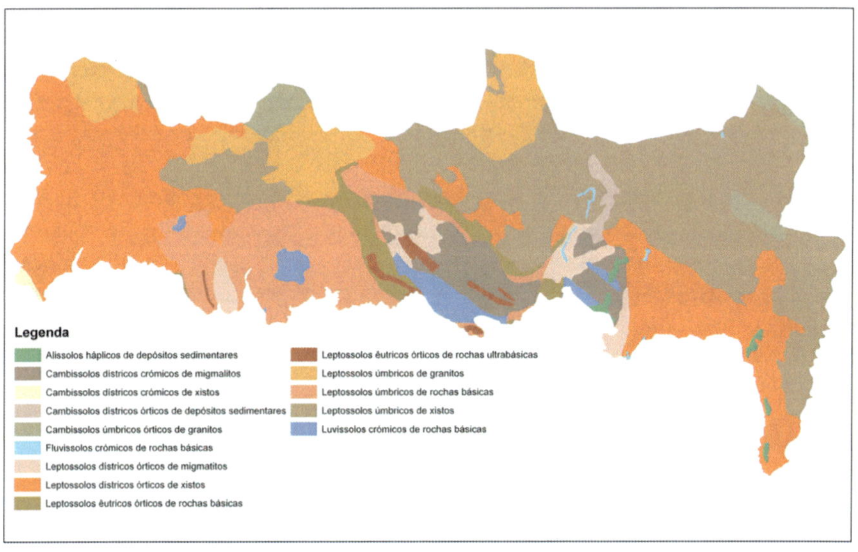

Soil Map of PNM: Main Soil Units

According to the Soil Map of Northeast Portugal, 1:100,000 (Agroconsultores e Coba, 1991), which follows the FAO system of classification (FAO/UNESCO 1987), Leptosols are largely dominant in PNM (77% of the area. These are shallow stony soils with hard rock at less than 0.5 m depth, occurring in steep slopes and convex landforms, which reflect the incipient pedogenesis prevailing under such topographic conditions, much more prone to morphogenetic activity. The Cambisols rank second in area (20%), covering the slopes lower than 12%, where deeper but still poorly developed soils form over any parent material. The more developed soils are Luvisols and Alisols, covering 2% of the territory (1.2 and 0.8%, respectively). They are both confined to the gentle slope plateau landforms, the first one developed over basic rocks, the second one over sedimentary deposits of Tertiary clayey material with rounded gravel. Other relevant soil units are the Fluvisols, developing along the few wider valley bottoms, over alluvial parent material. Regosols, above colluvial deposits, also occur in small patches, scattered all over the area. The former are deeper than the latter, they both make up the large group of incipient soils present in PNM.

In nearly half of the PNM the soils have a carbon rich surface (A) horizon (hence qualifying as umbric at secondary level). These are in well vegetated elevated areas, roughly above 900 m, where colder and wetter climatic conditions enhance the accumulation of organic matter. Soils are acid or very acid in most of the PNM area, either due to high organic content and or to parent material characteristics. In fact, soil units qualifying as eutric, which are non acid, cover only 4% of the area, strictly over basic lithology. Soils on the very steep slopes, in less than one tenth of the Park, qualify as lithic because hard rock is at very shallow depth (less than 0.1 m).

Soil properties reflect the processes and factors acting during soil formation and development. Although pedogenetic processes show incipient results, the influence of soil formation factors is clearly evident in major features of regional soils; this is true of PNM (Figueiredo, 2001). These effects have been implicitly mentioned, in part, above: the effect of climate on soil organic matter; of parent material on soil chemistry; of topography on soil depth. Other examples follow. Parent material also affects soil texture, since granite-derived soils are rich in coarse sand, those formed over schist are silt-rich, and those of basic rocks and sedimentary deposits have more clay. The combined effect of climate and topography on the content of rock fragments in the soil explains the presence of stony soils on gentle slopes in the drier areas; these are rarer in the wetter ones. Vegetation and land use affect carbon content. This is 3.5–4.5% on average in the A horizon of soils under forest, pasture or shrub cover, against 1% under annual and permanent crops. Even though this is an indication of the widely accepted detrimental effect of land use in soil resources, as explained later in this paper, Montesinho depicts striking examples of balanced farming systems in humanized landscapes, coming from traditional expertise in soil quality assessment and soil conservation principles based on good management practices.

Man and Land Use

The PNM includes 92 small villages inhabited by less than 8,000 people. Demographically the western part of PNM, although less populated, has many villages, most of them with less than 200 inhabitants; the eastern part villages are fewer in number but more populated, frequently over 300 inhabitants. Economically the PNM still produces grains, chestnuts and livestock, especially cattle and sheep. In recent years agriculture has declined, abandonment of the villages persists and the population is ageing. Most of the villages have too few children to keep the rural schools open. But it has been the traditional family livelihood of PNM inhabitants, which is based on small farming and sheep herding, that contributed definitely to the conservation of biodiversity in the region. For centuries, more than 50% of the territory was communal land, an ancestral Portuguese and European system of land property and management. Most of that land was privatized or taken by the State in the nineteenth and twentieth Centuries and today only one third of PNM remains communally owned (IPB/ICN, 2006).

The western part of the Montesinho Natural Park lies in the northern half of Vinhais municipality, where good examples of rural and agricultural lifestyles combined with wildlife preservation are found. At present, sweet chestnut *Castanea sativa* plantations are the main driver of economic activity in Vinhais. Cattle husbandry of the local breed *Mirandesa* is also economically important but unfortunately it is declining. Upland meadows, marked by a good network of hedgerows dominated by ash *Fraxinus angustifolia*, elm *Ulmus procera*, alder *alnus glutinosa* and poplars *Populus* sp., together with heath lands and cropland, compose a very green landscape, typically associated with *Mirandesa* husbandry. Recently created is a Biological Park ("Parque Biológico de Vinhais"), in Cidadelhe Mountai, which aims to generate a better understanding of wildlife in this area.

Chestnut trees (*Castanea sativa*) in Vinhais area, economically a very important product; the fruits also form part of the PNM logo (photo by A. Borges)

"Mirandesa" autochthonous cattle breed: grazing meadow in Babe (photo by S. Nobre)

The eastern part of the Montesinho Natural Park corresponds to the northern part Bragança municipality. Here, forestry, sheep herding and agriculture also combine with wildlife preservation. The Natural Park took its name from the most northern mountain of Bragança municipality that reaches an altitude of 1,487 m. The vegetation is composed of scrubs of *Erica* spp., *Genista* spp., *Chamaespartium* sp. and *Salix* spp. *Montesinho* is the only village at such heights; agricultural activity has decayed over approximately 20 years but tourism and recreational activities have became more and more significant. For most of the year there are only about 200 sheep grazing; however, grazing can be very intense from May to August when about 5,000 sheep are transported from the surrounding lowlands to graze in the highlands during summer months. Consequently, this area is very often subjected to wild fires that are mainly induced by shepherds who wish to obtain better grazing (Geraldes and Boavida, 2003). *Serra Serrada* reservoir, built to supply water to the city of Bragança and to generate hydroelectric power, is located in this mountain.

Alta Lombada seen from Bragança: the eastern plateau and drier tract in PNM (photo by A. Carvalho)

Major Natural Values

The main native forest cover was oak (*Quercus pyrenaica*) and chestnut trees (*Castanea sativa*). After centuries of destruction, wildlife populations are now recovering in the region in response to lessening of human pressure coupled with the expansion of scrub areas, better protection laws, the positive action of the Park management authority and a slow change in the attitude of individuals towards wildlife. Biodiversity is currently high but the densities of wild populations are lower than in the past (some, such as bear and lynx, reduced to extinction), mainly because man has destroyed most of the native forests. At the beginning of the twentieth Century, deer were exterminated, but by 1989, they came back via populations in Spain (Alves, 2004).

Oak trees (*Quercus pyrenaica*) in winter: the original forest cover and a wildlife habitat rated as excellent natural value (photo by S. Nobre)

This Park has a high flora and fauna diversity and the latter expressively contributes to its emblematic biodiversity. All of the trophic levels are present and large carnivores and wild and domestic herbivores coexist in the region. A population of about 30 Iberian wolves (*Canis lupus signatus* Cabrera) lives in the Park and it is linked to a larger population of more than 120 wolves in neighboring regions in Portugal and Spain. Other carnivores include common genet and red fox (*Vulpes* vulpes). Wild boar (*Sus scrofa*), roe deer (*Capreolus capreolus* Linnaeus), deer (*Cervus elaphus* Linnaeus) and, as mentioned above, domestic herbivores such as cattle, goat and sheep regional breeds, all inhabit the Natural Park. Main aquatic mammal species is the European otter (*Lutra lutra*) (IPB/ICN, 2006).

Dense riparian vegetation along the Maçãs river in Quintanilha, the eastern PNM border (photo by A. Carvalho)

Stream banks in PNM are covered by dense riparian vegetation dominated by alder *Alnus glutinosa*, ash *Fraxinus angustifolia*, poplar *Populus nigra* and willows *Salix* sp., limiting the primary production in these systems where the food webs are energetically dependent on allochthonous inputs of organic matter (Teixeira and Cortes, 2007).

These streams and their dense riparian galleries represent key features for the high level of biodiversity preservation in the PNM. They are subjected to a reduced human pressure, which contributes to low impact on water composition. The fish community is dominated by native brown trout *Salmo truta* populations, but endemic cyprinid species are also present, such as the Iberian chub *Squalius carolitertii*, the Iberian nase *Chondrostoma duriensis*, *Squalius alburnoides* and *Barbus bocagei*. The brown trout are present in the upper reaches of these rivers, whereas the cyprinids cohabit with trout, although in low densities, in the lower ones (Teixeira and Cortes, 2007). The water temperature ranges from a winter minimum of 4ºC to a summer maximum of 20ºC and their ecosystem function is, as already stressed, highly dependent of the input of allochthonous materials (Teixeira and Cortes, 2006).

Lower reaches include many old watermills, most of them abandoned or destroyed, but also with some good examples of structural and functional rehabilitation for tourism, recreation, and environmental and cultural education. They were essential to mill all cultivated cereals, mainly rye and wheat, the basis of human sustainability for centuries. The watermills required the presence of a succession of small artificial weirs creating the corresponding deep pools and riffles. The impact of human activities in these systems is limited because population is scarce and agriculture is extensive with low fertilizer input, thus contributing to the good water quality found all along the streams. However, an increase in fishing pressure, often using illegal procedures (poison, nets), is responsible for the gradual decrease of natural trout stocks and has lead local authorities to follow active management programmes, such as the implementation of stocking operations.

Remnant woodlands of Holm oak (*Quercus ilex)* represent also important spots of diversity in the PNM, in spite of covering only around 3,000 ha. Usually they are small in size and simple in shape, and occur close to each other, frequently located towards the bottom of very steep slopes, facing West Northwest and East Southeast, and relatively close to ephemeral streams. However, these woodlands intercept these streams in most of the cases. Their patterns do not change when woodlands are considered according to their development stage (Dias and Azevedo, 2008). Most

Ultramafic flora, as *Alyssum pintodasilvae* and *Santolina semidentata*, are qualified as excellent natural value, thus requiring conservation awareness (photo by C. Aguiar)

of these holm oak remnant woodlands are over mafic and ultramafic rocks derived soils that are frequent in this territory and have very particularly vascular indigenous or synantropic plant *taxa* (Aguiar, 2002, see also article by Sequeira et al. in this chapter).

Productive forests were introduced during the twentieth century and most of them are pure stands of several species, the most important being *Pinus pinaster* Ait. and *P. nigra* Arn., and the less represented *P. sylvestris* L. and *P. strobus* L. Every year those pine stands are severely attacked by *T. pityocampa*, a forest pest that field observations and experimental confirmation suggested preference for this host (Arnaldo and Torres, 2006). These plantations were promoted by the National Forest Service and included also *Pseudotsuga menziesii* as well as some deciduous species, *Betula alba* and *Quercus robur*, which are common in the highlands of Montesinho and Coroa mountains. Stands have approximately uniform tree density and natural thinning, natural regeneration and lack of management have created clearings of different sizes. A shrub understorey is composed dominantly by *Erica* spp., *Ulex* spp., *Cistus* spp. and *Cytisus* spp.

These plantations are common in the eastern part of PNM, where they have been an important winter habitat used both by red and roe deer. Distributions of red and roe deer populations overlap and have been recently expanding in PNM, partially due to favorable habitat changes and limited culling policies adopted by the Park management authority. Estimated densities of red and roe deer in the area are approximately of 0.03–0.04 and 0.01–0.02 individuals/hectares, respectively (Ramos et al., 2006). Deer rubbing of adult trees is thought to be an important source of economic losses in the area. Damage to trees, in orchards and agricultural crops, has been increasingly reported by farmers and foresters inhabiting the area.

Fauna and Flora Values Formally Recognized in the PNM

The Habitats Directive (Council Directive 92/43/EEC) and Birds Directive (Council Directive 79/409/EEC) form the cornerstone of Europe';s nature conservation policy. It is built around two pillars: the Natura 2000 network of protected sites and the strict system of species protection. The high value of biodiversity of Montesinho Natural Park gives it this special statute at international level.

A harrier (*Circus pygargus*) and a dove (*Streptopelia turtur*) are the highest naturalistically valued birds of PNM. Three mammals – the Iberian wolf (*Canis lupus*), a mole (*Galemys pyrenaicus*) and an otter (*Lutra lutra*) -, one reptile – a turtle (*Mauremys leprosa*) – and an invertebrate – a freshwater pearl mussel (*Margaritifera margaritifera*) – are also considered excellent natural values of PNM Fauna. Many flowering plants that occur in PNM are classified as excellent value such as *Dianthus marizii*, *Festuca brigantina*, *F. elegans*, *F. summilusitanica*, *Jasione crispa* ssp. *Serpentinica*, *Santolina semidentata*, and *Veronica micrantha*.

Meadow along a meandering stream with riparian corridor, near Babe (photo by A. Carvalho)

Many other species are classified as of good nature value such as the birds – *Anthus campestris*, *A. spinoletta*, *A. trivialis*, *Aquila chrysaetos*, *Asio flammeus*, *Bubo bubo*, *Caprimulgus europaeus*, *Ciconia nigra*, *Circus cyaneus*, *Emberiza hortulana*, *Falco peregrinus*, *Lanius collurio*, *Monticola saxatilis*, *Oenanthe hispanica*, *O. oenanthe*, *Pernis apivorus*, *Phoenicurus phoenicurus*, *Sylvia borin* -, the fishes – *Chondrostoma polylepis* and *Rutilus alburnoides* -, the invertebrate – *Unio crassus* – and the flowering plant – *Narcissus asturiensis*.

Natural and semi-natural habitats mixed in the PNM landscape gradients are essential to maintain all of these species. The habitats recognized as having global excellent value are the dry heaths and the oak woods with *Quercus pyrenaica*; many other habitats are recognized as of global good value such as the *Castanea sativa* woods, the endemic heaths with gorse, the hydrophilous tall herb fringe communities of plains and mountain, the lowland hay meadows, the pseudo-steppe with grasses and annuals, the *Quercus ilex* and *Quercus rotundifolia* forests, the rivers with muddy banks, the siliceous rocky slopes with chasmophytic vegetation, the temperate Atlantic wet heaths, and the water courses at all levels, from the mountains to the plain (http://eunis.eea.europa.eu/).

Photos By

A. Carvalho, IPB / ESA / CIMO, anacarv@ipb.pt
A. Borges, Arborea, Vinhais, aborges@arborea.pt
A. Suarez, IPB, atilano@ipb.pt

References

Agroconsultores e Coba. 1991. Carta dos Solos, Carta do Uso Actual da Terra e Carta da Aptidão da Terra do Nordeste de Portugal. Vila Real: UTAD/PDRITM.

Aguiar, C. 2002. Flora e Vegetação da Serra da Nogueira e do Parque Natural de Montesinho. Instituto Superior de Agronomia. Lisboa: Universidade Técnica de Lisboa: 661.

Alves, J. 2004. Man and wild boar: a study in Montesinho Natural Park, Portugal. In Galemys: Boletín informativo de la Sociedad Española para la conservación y estudio de los mamíferos 16 (1): 223–230.

Arnaldo, P. & Torres, L. 2006. Effect of Different Hosts on Thaumetopoea pityocampa Populations in Northeast Portugal. In Phytoparasitica 34 (5): 523.

Dias, R. & Azevedo, J. C. 2008. Distribution and spatial configuration of holm oak woodlands in the Montesinho/Nogueira site, Portugal. In Landscape Ecology and Forest Management – Challenges and Solutions. Proceedings of the International Conference IUFRO – 8.01.02 Landscape Ecology. Chengdu, China.

FAO/UNESCO. 1987. Soil Map of the World, Revised Legend, Amended Fourth Draft. Rome: United Nations Food and Agriculture Organization.

Ferreira, A. de B. (coord.) 2005. Geografia de Portugal: I – O Ambiente Físico (dir. C. A. Medeiros). Mem Martins: Círculo de Leitores.

Figueiredo, T. de. 1990. Aplicação da Equação Universal de Perda de Solo na estimativa da Erosão Potencial: o caso do Parque Natural de Montesinho. Bragança: Instituto Politécnico de Bragança: 87.

Figueiredo, T. de. 2001. Recursos pedológicos do Nordeste Transmontano. In II Seminário sobre Recursos Naturais do Nordeste Transmontano. Bragança: Instituto Politécnico de Bragança: 15.

Figueiredo, T. de & Fonseca, F. 1997. Les sols, les processus d'érosion et l'utilisation de la terre en montagne au Nord-Est du Portugal: Approche cartographique sur quelques zones du Parc Naturel de Montesinho. In Réseau Erosion 17: 205–217.

Geraldes, A. M. & Boavida, M. J. 2003. Distinct age and landscape influence on two reservoirs under the same climate. In Hydrobiologia 504 (1): 277–288.

Gonçalves, D. A. 1980. Parque Natural de Montesinho. Lisboa: Serviço Nacional de Parques, Reservas e Património Paisagístico.

Gonçalves, D. A. 1985. Contribuição para o estudo do clima da bacia superior do rio Sabor (Influência da circulação geral e regional na estrutura da baixa atmosfera). Vila Real: Instituto Universitário de Trás-os-Montes e Alto e Douro.

Gonçalves, D. A. 1991a. O clima e os ecossistemas Agro-Ecológicos do Parque Natural de Montesinho. In II Seminário Técnico sobre Conservação da Natureza nos Países do Sul da Europa, 4ª Secção – Climatologia. Faro: 18.

Gonçalves, D. A. 1991b. Terra Fria – Terra Quente (1ª Aproximação). Bragança: IPB/Centro de Agroclimatologia da UTAD.

Henriques, A. Gonçalves. 1985. Avaliação dos Recursos Hídricos de Portugal Continental: Contribuição para o Ordenamento do Território. Lisboa: Instituto de Estudos para o Desenvolvimento.

IPB/ICN. 2006. Plano de Ordenamento do Parque Natural de Montesinho: I – Relatório de Caracterização. Bragança: Instituto Politécnico de Bragança.

Medeiros, C. A. 1987. Introdução à Geografia de Portugal. Lisboa: Imprensa Universitária, Editorial Estampa.

Pereira, P., Pereira, D. I., Caetano-Alves, M. I. & Meireles, C. 2003. Geomorfologia do Parque Natural de Montesinho: controlo estrutural e superfícies de aplanamento. In Ciências da Terra (UNL) Special issue V: C61–C64.

Ramos, J., M. Bugalho, Cortez, P. & Iason, G. R. 2006. Selection of trees for rubbing by red and roe deer in forest plantations. In Forest Ecology and Management 222 (1–3): 39–45.

Ribeiro, A. C. 1996. Análise da ocorrência de geadas e estimativa da temperatura mínima na relva em condições de geada de radiação: estudo na bacia superior do rio Sabor. Vila Real: UTAD.

Ribeiro, O., Lautensach, H. & Daveau, S. 1987. Geografia de Portugal: I – A Posição geográfica e o Território. Lisboa: Edições João Sá da Costa.

Taborda, V. 1987. Alto Trás-os-Montes: Estudo Geográfico. Lisboa: Livros Horizonte (reed. Universidade de Coimbra, 1932).

Teixeira, A. & Cortes, R. M. V. 2006. Diet of stocked and wild trout, Salmo trutta: Is there competition for resources?. In Folia Zool 55 (1): 61–73.

Teixeira, A. & Cortes, R. M. V. 2007. PIT telemetry as a method to study the habitat requirements of fish populations: application to native and stocked trout movements. In Hydrobiologia 582 (1): 171–185.

The Geological Heritage of Montesinho Natural Park (Portugal)

Carlos Meireles

Somewhere between 400 and 300 Ma ago, a complex collisional plate tectonics process took place involving three continents, Gondwana, Laurentia and Baltica.

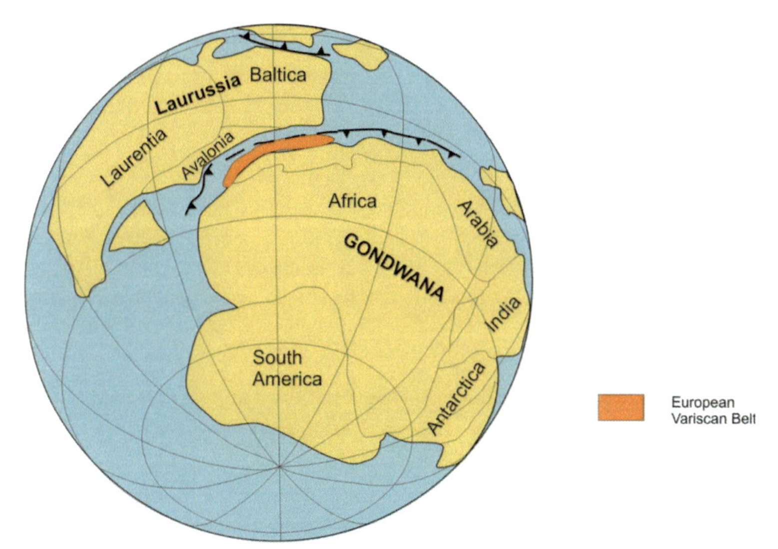

Paleogeographical reconstruction of continental masses during Early Carboniferous, 350Ma (adapted from Winchester et al. 2002)

In this process, part of the crustal materials, mainly oceanic crust, plunged below the less dense continental crust, in a subduction zone. Another part was obducted and thrust over the continental mass. The latter units are termed allochthonous, in contrast with those keeping their original emplacement, termed autochthonous. The allochthonous pile of rocks moved over the continent for hundreds of kilometers. From this assembling, a supercontinent was born: Pangea.

C. Meireles (✉)
Departamento de Geologia, Instituto Nacional de Engenharia, Tecnologia e Inovação, S. Mamede de Infesta, Portugal
e-mail: Carlos.Meireles@ineti.pt

N. Evelpidou et al. (eds.), *Natural Heritage from East to West*,
DOI 10.1007/978-3-642-01577-9_16, © Springer-Verlag Berlin Heidelberg 2010

Different orogenic mountain belts resulted from that collision. In Europe, the event is known as variscan orogeny and the subsequent mountain belt as variscan chain.

In the Iberian Peninsula the variscan chain is well preserved, as the Iberian Massif. According to paleogeographical, lithological, tectonic and metamorphic evidences, the Iberian Massif is subdivided in different zones.

Our focus is on the north-western part of the Iberian Massif, where the Galicia – Trás-os-Montes Zone (GTMZ) occurs. It is mainly formed by allochthonous and para-autochthonous units bounded by thrusts planes. All together, these tectonic units are stacked upon the continental margin of Gondwana autochthon sedimentary sequence (Central Iberian Zone). As expected, this fragmented stacked pile of rocks is a puzzling challenge for geologists attempting to reconstruct Earth's history. The main feature of this zone is the presence of the so called mafic and ultramafic massifs. These "exotic" massifs exist in complex imbrications of several tectonic units from different origins: continental crust and mantle rocks (relict granulites, eclogites and gneisses) assembled together with basic and ultra basic rocks, testifying a probable old oceanic crust (± 1000 Ma) involved in the variscan orogeny. These tectonic slides are thrust over fragmented Paleozoic ophiolite and several types of metasediments.

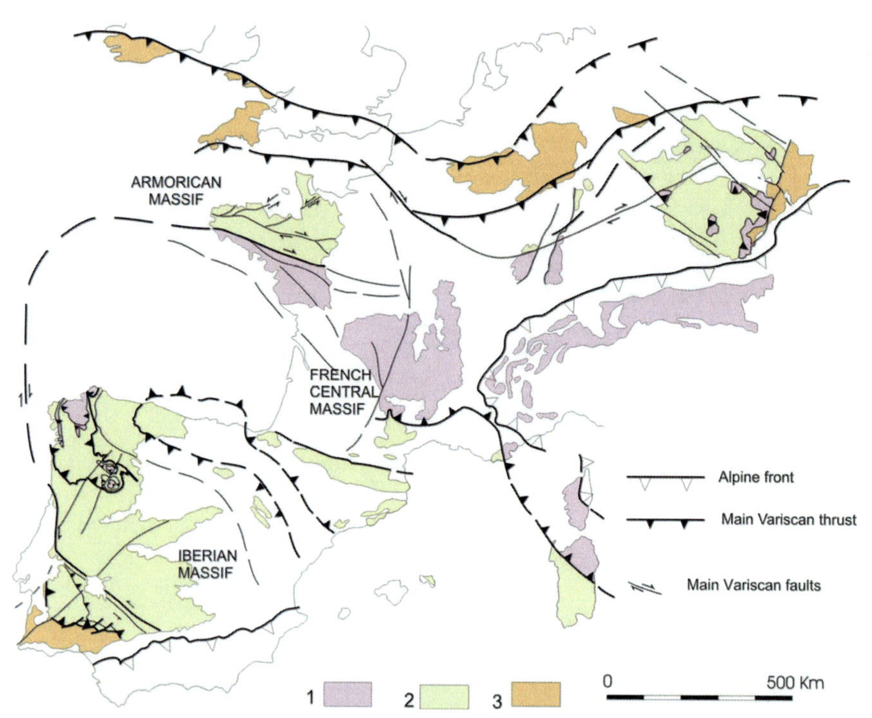

Geological sketch map of the European Variscan Orogeny (after Martínez-Catalán 1990). 1 – Allochtonous Terranes; 2 – Autochtonous and parautochtonous Terranes with Gondwanic affinity; 3 – Original thrust belt and foredeep basins

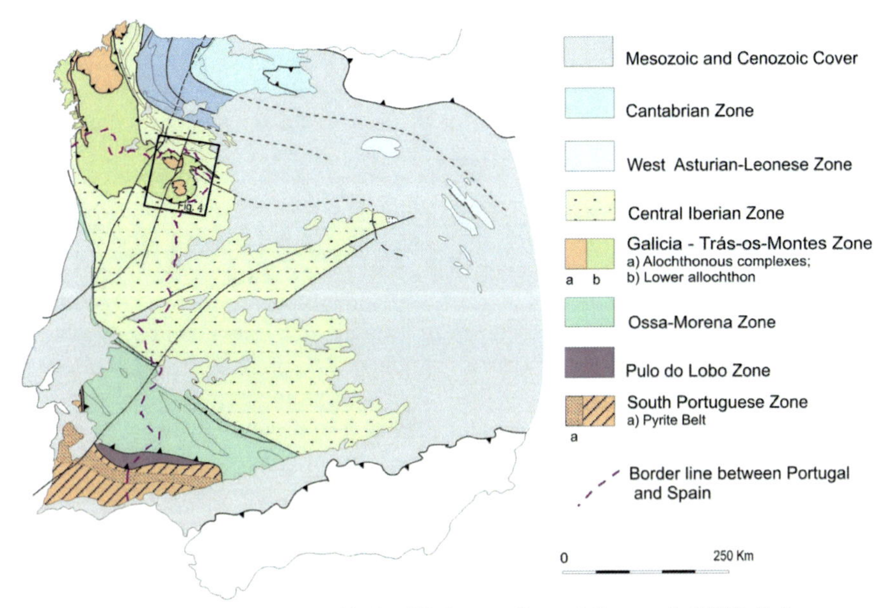

Subdivisions of the Iberian Massif (simplified), according to Julivert et al. (1972), Farías et al. (1987) and González Clavijo (1997)

The Bragança Massif, one of these exotic terrains, occurs near Montesinho Natural Park (PNM) and it is one of the main geological features of this region. It comprises mainly two NW-SE bodies with a broad lense-like shape. It is composed of three main allochthonous units stacked over a basal para-autochthonous

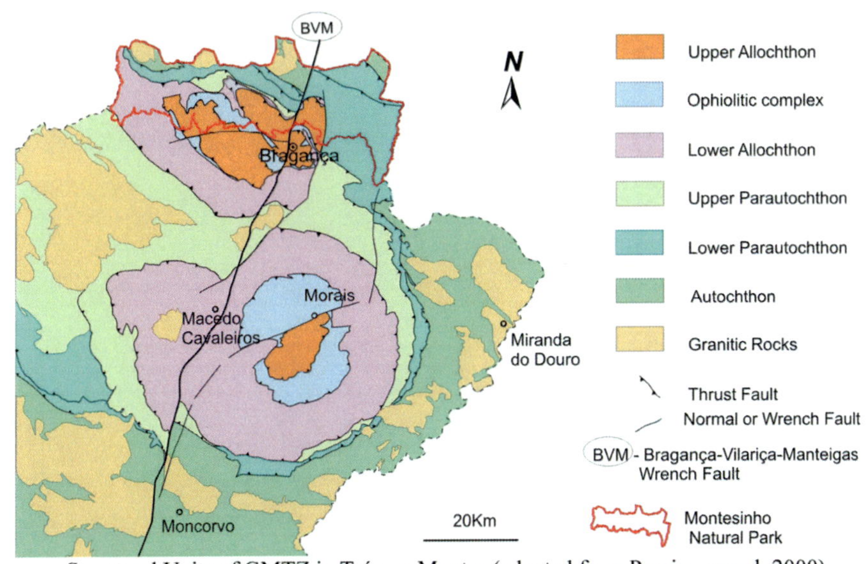

Structural Units of GMTZ in Trás-os-Montes (adapted from Pereira, coord. 2000)

thrust complex, which includes metasediments of Silurian to Devonian age, affected by greenschist facies and having lithological and paleogeographic affinities with the subautochthon and autochthon units (Farias et al., 1987; Iglésias et al., 1983; Ribeiro, 1974; Ribeiro et al., 1990). From the base to the top, the pile is formed by:

- A Lower Allochthonous Thrust Complex (LATC), characterized by the presence of bimodal magmatism (Ribeiro, 1991 and references therein) and composed of metasediments ascribed to Ordovician/Lower Silurian age at the base and Silurian/Lower Devonian at the top (Ribeiro et al., 1990).
- An Ophiolite Thrust Complex (OTC), with amphibolites, gabbros, serpentinites and grey phyllites that underwent strong tectonic disruption and retrograded to greenschist facies during the variscan orogeny. The age is Devonian (Dallmeyer and Gil Ibarguchi, 1990).
- An Upper Allochthonous Thrust Complex (UATC) mainly composed of high-grade metamorphic mafic granulites, gneiss with retrograded boudins of eclogites and metaperidotites hosting chromitites. These ultramafic rocks most likely represent either the remnants of supra-subduction zone (SSZ), magmatism in a SSZ spreading centre above a depleted mantle (Bridges et al., 1995) or an arc crust/mantle interface during subduction zone evolution (Moreno et al., 1999; Gil Ibarguchi et al., 1999).

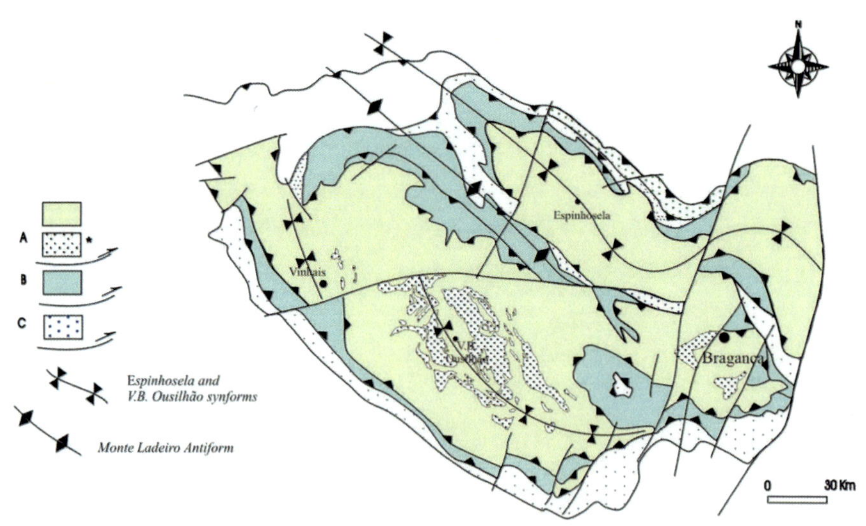

Geological outline of the Bragança massif. (A) – Allochthonous Thrust Complex (UTC) * ultramafic outcrops; (B) – Ophiolite Thrust Complex (OTC); (C) – Lower Allochthonous Thrust Complex (LATC). Adapted from Ribeiro (1974)

Hence, most of the area of Montesinho Natural Park is occupied by the parautochthonous and autochthonous sequences. The Parautochthonous Complex consists of Paleozoic lithologies, Silurian – Devonian ages, in a tectonically imbricate thrust system, mainly composed of phyllites with several lenses of black cherts, quartzites, limestones and metavolcanic rocks. Sandstones and shales, also represented, are Upper Devonian age.

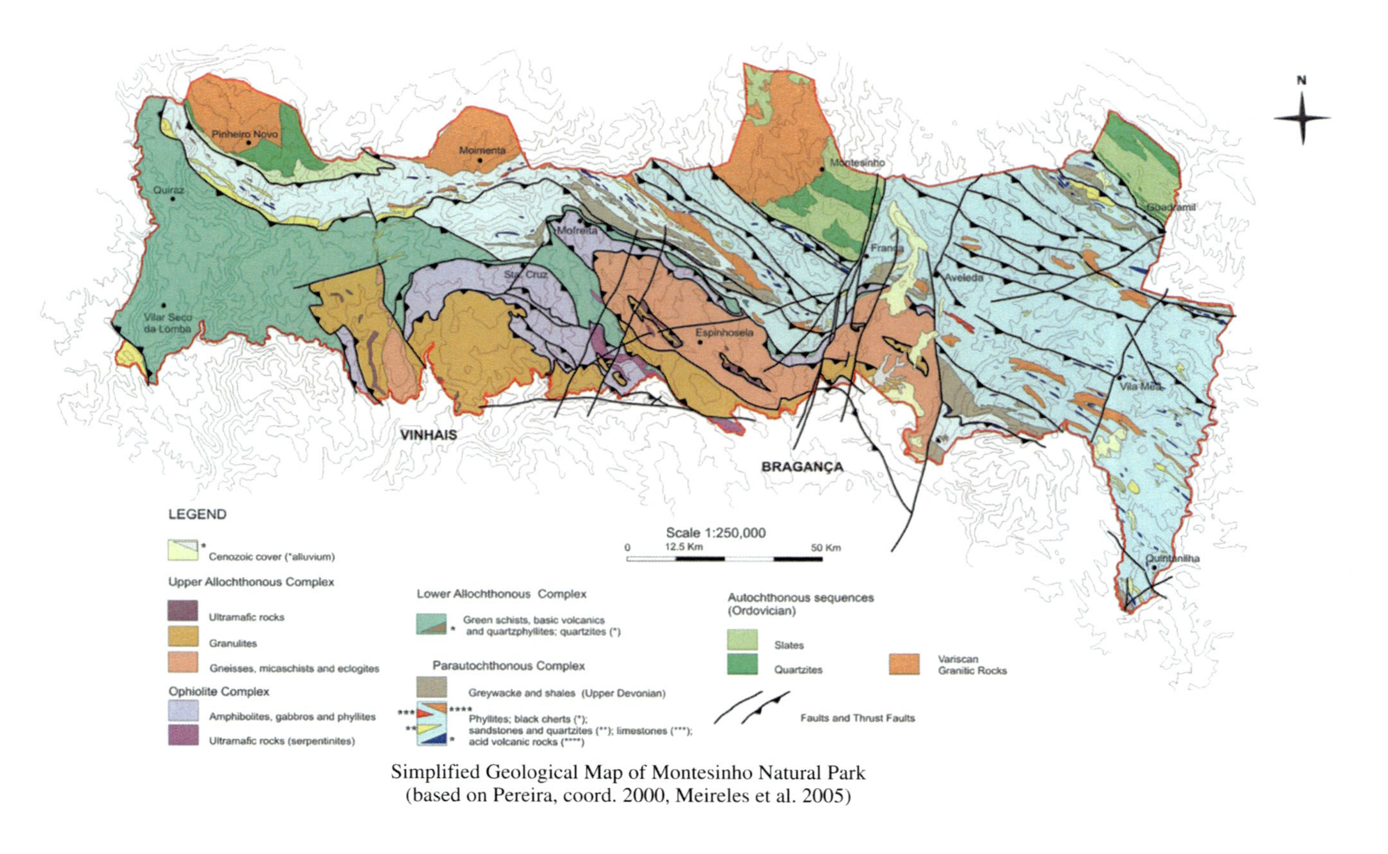

Simplified Geological Map of Montesinho Natural Park (based on Pereira, coord. 2000, Meireles et al. 2005)

The autochthonous sequence consists of slates and quartzites, representing the complete range of the Ordovician (480–444 Ma). They occur near Rio de Onor, Guadramil, França and Pinheiro Velho villages. The quartzites levels are rich in fossils widely known in the Lower Ordovician facies of SW Europe, namely brachiopods and bivalves, and specially ichnofossils (Sá, 2006; Sá et al., 2005). Close to the Spanish border, near Rio de Onor (NE PNM), there is a place known in the folk traditions as Pedras Escrevidas ("Written Stones"). Although initially interpreted by archaeologists as a pre-roman writing, the first geological studies of the area ascribed these features to algae remains (Meireles, 2000 and references therein). More recent geological mapping and studies allowed their interpretion correctly as ichnofossils of the genus and species *Daedalus halli* (Rouault) and *Daedalus labechei* (Rouault) (Meireles, 2000; Meireles and Sá, 2001; Sá et al., 2002). This spectacular paleontological phenomenon, without counterpart in the world, is currently under detailed study as it is a valuable part of the Iberian geological heritage (Sá et al., 2008). In the Upper Ordovician units, the widespread dropstone level (diamictites) is the evidence of a huge glaciation that had covered the whole Gondwana continent in the final stage of the Ordovician period.

Fusion of crustal materials occurred as the final stage of the above mentioned collision process between Gondwana and Laurasia. As a result, granitic magmas ascending to the upper levels of the crust created several granitic intrusions. In the PNM there are three granitic massifs, part of huge batholiths, continuing into Spain.

Then a new phase started for the Pangea continental mass. During the Mesozoic a strong erosional process took place, being responsible for flattening its morphology. The present day Iberian Meseta plateau, generally at 800 m above sea level, is the evidence of this process.

A new orogenic cycle (Alpine orogeny) started with the opening of the Atlantic Ocean (around 145 Ma), lasting until the present. A new relief was built, with the

"Pedras Escrevidas" or "Written Stones" outcrop: a detail of *Daeadalus halli* (Rouault) (hammer length is 28cm)

uplift and depression affecting the continental mass, along with faults. Hence the present landscape of Montesinho Natural Park is not only determined by the lithologies but also by the alpine faults (Meireles et al., 2002; Pereira et al., 2003). One of these, the so called Bragança – Vilariça – Manteigas fault, crosses the Park separating a western horst (uplift block) and an eastern graben (lower block) of the Baçal area. The remains of the Cenozoic cover testify to the more recent erosion processes, and they are formed by coarse and sandy alluvial deposits, occurring in the eastern block of this fault system. Several geomorphological domains can be defined and are controlled by: several lithologies, alpine faults, fault slopes and deep river valleys (Pereira et al., 2003).

Geological resources are produced by geological processes and, as expected, PNM is no exception. Several mining activities have taken place since Antiquity. Gold and tin were exploited by the Romans. In the twentieth century, Portelo was the most important Portuguese tin mine in operation. In the Ordovician quartzites, namely near Guadramil, strata bound iron levels occur, with ore reserves calculated to be almost 6 Mt (Meireles, 2000 and references therein). Massive chromitite bodies have been found in the Bragança Massif. An attempt to study and characterize these deposits was made by Neiva (1947 and references therein), who fist reported the presence of platinum-group minerals in the chromite occurrences. Recently, more studies of the PGE mineralization in the chromitite occurrences were undertaken (Bridges et al., 1993). The Morais Massif (60 km south of Bragança) has extensive ultramafic sequences, but no chromitite bodies have been reported. The reason is that, in this area, ultramafic rocks belong to the Devonian ophiolitic complex while the ultramafic suite of the Bragança Massif belongs to the upper allochthonous unit. Exploitation of talc and ornamental rocks occurs in the ultramafic rocks of the ophiolite suite.

The PNM complex multiform geology is a good illustration of the way in which geodiversity regulates its diversified landscape, the groundwater supplies, and particularly how it controls vegetation. The best examples are the outcrops of the ultramafic rocks (both from the UATC and OTC units). The mineralogy and the chemical composition of these rocks are rich in metals (namely Ni and Cr) whose anomalous concentration is toxic for the majority of plants. However, some species have developed specific features that allow them to live on such soils (see elsewhere in this chapter). This geological control is also very clear outside the ultramafic outcrops. There is a clear contrast in the vegetation cover between the lithologies of the Bragança Massif (gneiss and amphibolites) and the phyllites and slates of the parautochthonous and autochthonous units. The gneiss and the amphibolites of Espinhosela synform are a better geological reservoir of groundwater than the phyllites. There is a distinct mineralogical and chemical composition in these rocks, with a more diversified mineralogy in the massif, when compared to the phyllites. Hence, different soils are formed by rock weathering; in the gneiss and the amphibolites the soils are chemically rich in metal nutrients, whereas the soils of the phyllites are chemically poor. Another distinct situation is that of the granitic areas where, in addition to parent material, specific mineralogy and chemical composition, elevation plays an important role in the control of the vegetation cover.

References

Bridges, J. C., Prichard, H. M., Neary, C. R. & Meireles, C. A. 1993. Platinum-group element mineralization in the chromite-rich rocks of the Bragança Massif, northern Portugal. In *Transactions of the Institute of Mining and Metallurgy* Section B (102): 103–113.

Bridges, J. C., Prichard, H. M. & Meireles, C. 1995. Podiform chromitite-bearing ultrabasic rocks from the Bragança Massif, N Portugal: fragments of island arc mantle? In *Geological Magazine* 132: 39–49.

Dallmeyer, R. D. & Gil Ibarguchi, J. I. 1990. Age of the amphibolitic metamorphism in the ophiolitic unit of the Morais allochthon (Portugal): implications for early hercynian orogenesis in the Iberian Massif. In *Journal of the Geological Society of London* 147: 873–878.

Farias, P., Gallastegui, G., González Lodeiro, F., Marquínez, J., Martín Parra, L. M., Martínez Catalán, J. R., Pablo Maciá, J. G. & Rodríguez Fernández, L. R. 1987. Aportaciones al conocimento de la litoestratigrafia y estructura de Galicia Central. In *Museum Labarotary Marine Geology Fac. Ciências Univ. Porto Mem.* N 1: 411–431.

Gil Ibarguchi, J. I., Abalos, B., Azcarraga, J. & Puelles, P. 1999. Deformation, high-pressure metamorphism and exhumation of ultramafic rocks in a deep subduction/collision setting (Cabo Ortegal, NW Spain). In *Journal of Metamorphic Geology* 17: 747–764.

González Clavijo, E. J. 1997. *La geologia del sinforme de Alcañices, Oeste de Zamora.* Tesis Doctoral. Salamanca: Univ. Salamanca, Dep. Geología: 330 pp.

Iglésias, M. P. L., Ribeiro, M. L. & Ribeiro, A. 1983 La interpretation aloctonista de la estrutura del Noroeste Peninsular. In *Libro Jubilar J.M. Rios, Geologia de España*: 459–467. Madrid: Inst. Geol. Min. España.

Julivert, M., Fontbote, J. M., Ribeiro, A. & Conde, L. 1972. *Mapa Tectónico de la Península Ibérica y Baleares, 1:1.000.000: Memoria Explicativa 1974.* Madrid: Inst. Geol. Min. España: 113 pp.

Martínez-Catalán, J. R. 1990. A non-cylindrical model for the northwest Iberian allochthonous terranes and their equivalents in the Hercynian belt of Western Europe. In *Tectonophysics* 179: 253–272.

Meireles, C. 2000. *Carta Geológica de Portugal na Escala 1/50 000 e notícia explicativa da Folha 4-C (Deilão), 2ª Ed.* Lisboa: Instituto Geológico e Mineiro: 28 pp.

Meireles, C., Pereira, D. I., Alves, I. C. & Pereira, P. 2002. Interesse patrimonial dos aspectos geológicos e geomorfológicos da região de Aveleda – Baçal (Parque Natural de Montesinho, NE de Portugal). In *Comun. Inst. Geol. e Mineiro* t. 89: 225–238.

Meireles, C. & Sá, A. A. 2001. As Pedras Escrevidas do Alto do Martim Preto (Guadramil): mistério esclarecido. In *II Seminário de Recursos Naturais do Nordeste Transmontano.* Bragança: Escola Superior de Educação.

Moreno, T., Lunar, R., Phichard, H., Monterrubio, S. & Ortega, L. 1999. Mineralización de elementos del grupo del platino (EGP) en las cromititas de los macizos ultramáficos del Complejo de Cabo Ortegal. In *Geogaceta* 25: 135–138.

Neiva, J. M. Cotelo 1947. Platina no districto de Bragança. In *Est. Not. Trab. Serv. Fom. Min.* III (fasc. 1–2): 19–25.

Pereira, E. S. (coord.) 2000. *Carta Geológica de Portugal à escala 1:200.000, Folha 2.* Lisboa: Inst. Geol. Mineiro.

Pereira, P., Pereira, D. I., Caetano Alves, M. I. & Meireles, C. 2003. Geomorfologia do Parque Natural de Montesinho: controlo estrutural e superfícies de aplanamento. In *VI Cong. Nac. Geol., Ciências da Terra (UNL)*, Lisboa, n esp. V, CD-ROM: C61–C64.

Ribeiro, A. 1974. *Contribuition à l'étude Tectonique de Trás-os-Montes Oriental, Mem. n 24, Nova Série).* Lisboa: Serv. Geol. Portugal: 168 pp.

Ribeiro, A., Pereira, E. & Dias, R. 1990. Structure of the Centro-Iberian Allocthon in the northwest of the Iberian Peninsula. In R. D. Dallmeyer & E. Martinez-Garcia (eds.) *Pre-Mesozoic Geology of Iberia*: 220–236. Berlin: Springer Verlag.

Ribeiro, M. L. 1991. *Contribuição para o conhecimento estratigráfico e petrológico da região a SW de Macedo de Cavaleiros (Trás-os-Montes Oriental) (Mem. Nº 30).* Lisboa: Serv. Geol. Portugal: 106 pp.

Sá, A. A. 2006 Bioestratigrafia do Ordovícico do nordeste de Portugal (Zona Centro Ibérica): estado actual do conhecimento. In J. Mirão & A. Balbino (coord.) *VII Cong. Nac. Geologia*, Évora, Vol. II: 617–620.

Sá, A. A., Meireles, C. & Coke, C. 2002. Concentração maciça de *Daedalus labechei* (ROUALT) (icnofóssil ordovícico) no Alto do Martim Preto (Guadramil – Bragança): património paleontológico a preservar e divulgar. In *XVIII Jornadas de la Sociedad Española de Paleontologia e II Congresso Ibérico de Paleontologia*: 138–139. Salamanca: Univ. de Salamanca.

Sá, A. A., Meireles, C., Coke, C. & Gutiérrez-Marco, J. C. 2005. Unidades litoestratigráficas do Ordovícico da região de Trás-os-Montes (Zona Centro-Ibérica, Portugal). In *Comunicações Geológicas., I.N.E.T.I.* t. 92: 31–74.

Sá, A. A., Gutiérrez-Marco, J. C. & Meireles, C. 2008. The "written stones" of the Montesinho Natural Park: where palaeontology meets popular legend. In G. Brown (ed.) *Geology of the planet, 33rd International Geological Congress*, Oslo, Abstracts CD-Rom X-CD Technologies Inc., Norway, and www.33igc.org (File 1345366.html)

Winchester, J. A., Pharaoh, T. C. & Verniers, J. 2002. Palaeozoic amalgamation of Central Europe: an introduction and synthesis of new results from recent geological and geophysical investigations. In J. A. Winchester, T. C. Pharaoh & J. Verniers (eds.) *Palaeozoic Amalgamation in Central Europe (Special Publication 201)*: 1–18. London: Geological Society.

Ultramafics of Bragança Massif: Soils, Flora and Vegetation

Eugénio Sequeira, Calos Aguiar and Carlos Meireles

Introduction

The presence of allochthonous ultramafic rocks is one of the main characteristics of the geology of NE Portugal (see paper by Meireles in this chapter). These ultramafic rocks are scattered in two large massifs: Bragança and Morais. The altitude of the northern Bragança Massif ranges between 600 and 1,060 m; it is all included in a rainy and cold supramediterranean climatic belt. A significant area of the Bragança Massif is part of the Montesinho Natural Park. The Morais Massif is mainly mesomediterranean, with altitudes between 300 and 900 m. In the latter massif it usually rains less than 800 mm per year and the mean annual temperature is higher than 12°C.

The supramediterranean and the mesomediteranean bioclimatic belts are closely correlated with natural potential vegetation (climax vegetation) and land use. In the supramediterranean areas deciduous *Quercus pyrenaica* forests dominate, together with heathlands and traditional agricultural systems based on cattle raising, chestnut orchards and rye and potato crops. In the mesomediterranean belt the landscape is composed of small horticulture areas, olive and almond groves and vineyards, intermingled with evergreen *Quercus* forests and *Cistus* shrublands. In the regional folk classification of climate, these two bioclimatic belts are known, respectively, as "cold land" and "hot land".

The mineralogy and the chemical composition of ultramafic rocks are rather unusual and have a strong impact on soil genesis, plant evolution and the vegetation assemblage. Besides their significance from the point of view of geodiversity (see article by Meireles in this chapter), ultramafic areas have an enormous societal and scientific importance, due to their soil diversity and to their role as plant biodiversity refuges.

E. Sequeira (✉)
Liga para a Protecção da Natureza, Lisboa, Portugal
e-mail: eugenio.sequeira@sapo.pt

N. Evelpidou et al. (eds.), *Natural Heritage from East to West*,
DOI 10.1007/978-3-642-01577-9_17,

Soils

From different soil genesis conditions – topography (runoff, erosion and infiltration), macromorphology of the rock, climate (varying from the "cold land" with more than 600 mm water surplus and moderate water deficit, to the "hot land" with less than 200 mm water surplus and high water deficit) and vegetation cover – the following soil successions may be observed in the north-eastern Portugal ultramafic rocks (Sequeira and Pinto da Silva, 1992):

- Fine earth accumulation on rock crevices
- Lithic Leptosols, mostly on steep slopes
- Mollic Cambisols, mostly in gentle slopes in the "cold land"
- Chromic Luvisols in climax conditions in the "hot land"
- Gleyic Fluvissols, in poorly drained depressions where colluviation occurs

Vegetation contrast of an ultramafic outcrop with the surroundings (Bragança Massif). Historical photo by E. Sequeira (1981) showing A.R. Pinto da Silva (far left), the late Portuguese botanist who studied the flora and vegetation of ultramafic areas

The ultramafic rock minerals, more or less serpentinized (chrysotile, serpophite, antigorite, with or without chlorite and bastite), weather very easily. The weathering of these rocks corresponds to dissolution (Pedro and Bitar, 1966; Sequeira, 1969; Sequeira and Pinto da Silva, 1992) and leaching depletes dissolved components out of the soil profile. As mineral matter left in place is only a non-dissolved residual, soil mass can be about 5% of the original rock mass (Sequeira, 1969). This is one of the main reasons why ultramafic soils tend to be shallow.

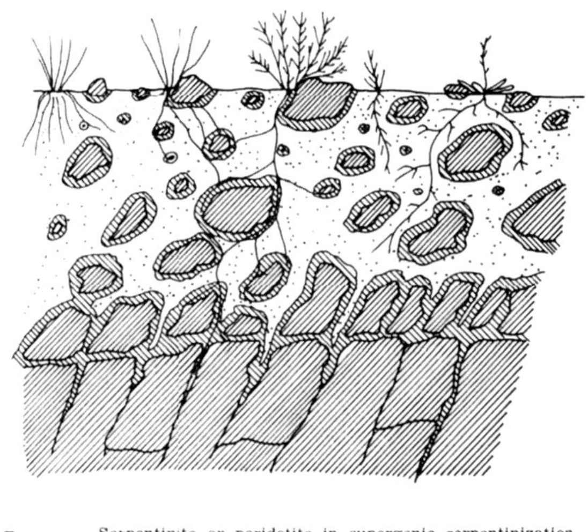

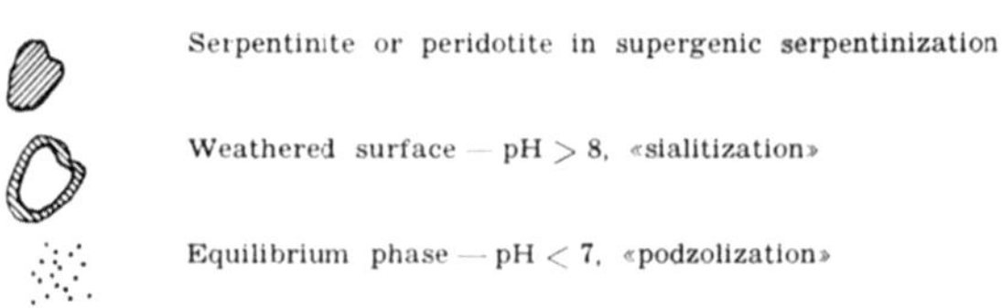

Schematic interpretation of the soil forming process of a Leptosol from ultramafic rocks (Sequeira & Pinto da Silva 1992)

Leaching rates depend on water surplus. Thus, in the "cold land" soils are shallow, whereas in hot and dryer places soils are deeper; this is a pattern very similar to that recorded in karstic zones. The order of leaching, expressed in oxides, is $MgO > NiO > SiO_2 > Na_2O > Cr_2O_3 > Fe_2O_3 > Al_2O_3$. In non-eroded soils with well preserved vegetation cover there is a biological accumulation of K, Ca, Cu and Zn in the top soils. This interpretation must be considered *sensu lato* because, under Mediterranean climate, the serpentinized rocks are very irregularly weathered, breaking down in blocks of different sizes that tend to become progressively smaller. So, different phases exists in the same soil horizon, especially when there is a high percentage of coarse fragments, as it is proved by significant variations of the pH within the same horizon: 8.6 or more in the weathered surface of rock fragments, about 7.0 in the soil grain size fraction between 0.4 and 2.0 mm, and less than 6.5 in the fraction less than 0.4 mm. This explains the simultaneous and intense leaching of Fe_2O_3, Al_2O_3, SiO_2, and MgO, and the prevalence of at least two different weathering processes at the same horizon (sialization and podzolization).

Flora and Vegetation

Toxicity to vascular flora is a well-known property of serpentine soils. The main adverse factors conditioning plant life in serpentine soils are probably the very high Mg/Ca ratio, the very high Ni, and low available N, P, K and Ca (Kruckeberg, 1986; Brady et al., 2005). Besides the unfavorable chemical soil properties, the flora and vegetation is also controlled by other environmental conditions such as the dominance of shallow soils, high soil temperatures and severe summer water shortage (Sequeira and Pinto da Silva, 1992; Kruckberg, 2002). These extreme ecological conditions are very selective to the flora and result in a widespread ecotypic differentiation among generalist plant species, and in a high diversity of endemic species and biogeographical disjunctions, some of them with a relict character (Kruckberg, 1986). In spite of the significant endemic species richness, at least in NE Portugal, the local scale species richness (i.e. diversity) is lower in serpentines than in the neighbouring mafic or felsic rocks.

In serpentine soils, serpentinofuges are eliminated or their populations depressed by adverse soil properties, while plant populations adapted to serpentines – the serpentinophytes – have lower competitive abilities in acid and basic rocks (Kruckeberg, 1986; Brady et al., 2005). Several genetic based tolerance mechanisms to cope with the serpentine effect have been described. Hyper accumulation is a striking mechanism that is hypothesized to explain higher plant tolerance to heavy metal or trace elements toxicity (Brooks, 1987). The *Alyssum pintodasilvae* (= *A. serpillyfolium* subsp. *lusitanicum; Brassicaceae*) is a Portuguese (and possibly Iberian) endemic nickel hyper accumulator. Tillage disturbance increases soil erosion and toxicity, and boosts the establishment of the *Alyssum pintodasilvae* communities (Sequeira, 1969; Sequeira and Pinto da Silva, 1992). Soil toxicity enhancement is linked with the mixture of different soil phases and with an acceleration of soil weathering processes (Sequeira, 1969).

Alyssum pintodasilvae, an endemic nickel hyperaccumulator
(photo by C. Aguiar)

The Bragança Massif is more diverse in endemic species and in other species important for conservation than the Morais Massif because these plants tend to concentrate their occurrence at higher altitudes. There are not yet satisfactory explanations for this pattern but, indeed, they should be grounded in the biogeographic context of the Bragança Massif. It is very close to a more or less continuous mountain system, from the Spanish Cantabric Mountains southwards, that facilitated the migration of a mountain basophyllous-neutrophyllous flora.

The significant IUCN (International Union for Conservation of Nature) threat category labeling many of the serpentinophytes (Aguiar et al., 1999) is mainly due to habitat intrinsic rarity. In the last 10 years, there is no direct evidence of a noticeable decline in their area of occupancy of threatened plants and/or in the quality of their habitats. The lack of agricultural and forestry suitability of these soils certainly had a positive contribution to their flora conservation.

Although forests in Portuguese serpentine soils are poor in endemics and biogeographical disjunctions, their dominant tree and shrub species are closely related to the rock substrate. *Quercus rotundifolia* is the only climax tree occurring in north-eastern Portugal serpentines. Forests on nearby basic or acid rock, with flatter geomorphologic forms, are instead dominated by *Quercus pyrenaica* (above 650–750 m) or by *Quercus suber* (at lower altitudes).

Table 1 Endemic species and other rare plants of the ultramafic outcrops of Northeastern Portugal (Aguiar, unpubl.)

Species	Bragança Massif	Morais Massif	IUCN categories of threat
Endemic			(national scale)
Alyssum pintodasilvae (*Brassicaceae*)	X	X	LC
Anthyllis sampaioana (*Fabaceae*)	X	-	NT
Arenaria querioides subsp. *fontqueri* (Caryophyllaceae)	X	X	NT
Armeria eriophylla (*Plumbaginaceae*)	X	-	NT
Avenula pratensis subsp. *lusitanica* (*Poaceae*)	X	X	VU
Festuca brigantina (*Poaceae*)	X	-	VU
Non endemic in Portugal exclusive of ultramafic rocks (includes biogeographical disjunctions)			
Antirrhinum braun-blanquetii (*Plantaginaceae*)	X	X	CR
Armeria langei subsp. *daveaui* (*Plumbaginaceae*)	X	X	NT
Armeria langei subsp. *langei* (*Plumbaginaceae*)	-	X	NT
Asplenium adiantum-nigrum subsp. *corunnense* (*Aspleniaceae*)	X	X	VU
Astragalus incanus subsp. *nummularioides* (*Fabaceae*)	X	-	VU
Bromus squarrosus (*Poaceae*)	X	-	EN
Dianthus laricifolius subsp. *marizii* (*Caryophyllaceae*)	X	X	NT
Elymus hispidus subsp. *barbulatus* (*Poaceae*)	X	-	EX
Gagea pratensis (*Liliaceae*)	X	-	EN
Jasonia tuberosa (*Asteraceae*)	X	-	EN
Linaria aeruginea var. *simplex* (*Plantaginaceae*)	X	X	VU
Notholaena marantae subsp. *marantae* (*Pteridaceae*)	X	X	VU
Reseda virgata (*Resedaceae*)	X	-	NT
Santolina semidentata (*Asteraceae*)	X	X	LC
Saxifraga dichotoma subsp. *albarracinensis* (*Saxifragaceae*)	X	-	EN
Seseli montanum subsp. *peixoteanum* (*Apiaceae*)	X	X	LC
Silene legionensis (*Caryophyllaceae*)	X	-	VU
Ventenata dubia (*Poaceae*)	X	-	EX

International Union for Conservation of Nature, IUCN, categories of threat: Extinct, EX; Critically Endangered, CR; Endangered, EN; Vulnerable, VU; Near Threatened, NT: Least Concern, LC. x/- refers to occurrence/non-occurrence.

Tall shrubby areas and forests recover slowly after catastrophic disturbance events (e.g. wildfires) in serpentine soils and they are strongly depressed by extensive sheep and goat grazing. Ultramafic soils tend to be shallow, erosion prone and their toxicity to plants (serpentine effect) is correlated with soil genesis, since incipient soils (e.g. Lithic Leptosols, see above) are much more selective to plants than moderately or well developed soils (e.g. Cambisols and Luvisols). Consequently, vegetation succession is a slow process in ultramafic outcrops, which are usually covered by open pioneer vegetation mosaics, dominated by annual plant communities, perennial grasses and poorly developed short shrubby vegetation. These habitats mosaics, however, are fundamental to plant conservation because more than 75% of the species listed in Table 1 have a phytosociological optimum in pioneer plant communities or in rocky habitats.

Some endemic species of the serpentine outcrops of Northeastern Portugal: *Avenula pratensis* subsp. *lusitanica* (top left); *Arenaria querioides* subsp. *fontqueri* (top right); *Armeria eriophylla* (bottom) (photos by C. Aguiar)

References

Aguiar, C., Honrado, J. J., Sequeira, M., Jansen, J., Caldas, F. B., Almeida da Silva, R. M. & Seneca, A. 1999. *Plantas vasculares e briófitas raras e a proteger no Norte de Portugal Continental*. Lisboa: V Jornadas de Taxonomia.

Brady, K. U. B., Kruckeberg, A. R. & Bradshaw Jr., H. D. 2005. Evolutionary ecology of plant adaptation to serpentine soils. In *Ann. Rev. Ecol. Evol. Syst*. 36: 243–66.

Brooks, R. R. 1987. *Serpentine and its Vegetation: a Multidisciplinary Approach*. Portland: Dioscorides Press.

Kruckberg, A. R. 1986. An essay: The stimulus of unusual geologies for plant speciation. In *Syst. Bot.* 11: 455–463.
Kruckberg, A. R. 2002. *Geology and Plant Life*. Portland: University of Washington Press.
Pedro, G. & Bittar, K. E. 1966. Contribution à l'étude de la genese des sols hypermagnésienes: recherches experimentales sur l'alteration chimiques des roches ultrabasiques (serpentines). In *Annls Agron.* 17: 421–430.
Sequeira, E. M. & Pinto da Silva, A. R. 1992. Ecology of serpentinized areas of North-East Portugal. In B. A. Roberts & J. Proctor (eds.) *The Ecology of Areas with Serpentinized Roks. A World View*: 169–197. Netherlands: Kluwer Academic Publishers.
Sequeira, E. M. 1969. Toxicity and movement of heavy metals in serpentinic soils (North-Eastern Portugal). In *Agron. Lusit.* 30 (2): 115–154.
IUCN 2001. *IUCN Red List Categories and Criteria: Version 3.1.* IUCN Species Survival Commission. Gland, Switzerland and Cambridge, UK: IUCN.

Land Use, Landscape and Sustainability: Examples from Montesinho

J. Castro

The traditional and multifunctional landscapes of Montesinho Natural Park (PNM), with their typical complexes of agro-, silvo- and pastoral components, changed thoroughly during past decades. Historical, social, economic and cultural factors, such as poor communications, biophysical events, and direct contact with nature in everyday life should be taken into account to explain its present land use pattern.

The current land use patterns are based in an ancestral arrangement of factors resulting from a combination of two main parameters: water availability and village proximity, both of them highly dependent of the topographic circumstances. As a result, four main land use groups must be considered: vegetable gardens and orchards near village streams margins, mainly over Fluvisols; meadows along the streams, also over Fluvisols; open cereal fields around the village, frequently over Dystric Cambisols and Leptosols; and more or less wooded open land outside this agricultural matrix, on Umbric Leptosols (IPB/ICN, 2006). The last one, the open woodland matrix, is the largest component of PNM landscape. Essentially, it is an export ecosystem: shrubs, firewood, pasture, but also, rock outcrops, honey, etc. In contrast, the vegetable gardens and orchards benefit greatly from human and animal labor, manure and water, in order to produce the seasonal fresh food to complement the inhabitants' diet, which is mainly based on cereals and meat. The meadows are, or were, essentially a "power" ecosystem feeding cattle used to plough fields. The open cereal fields ultimately provide bread, the basis of human life.

J. Castro (✉)
Instituto Politécnico de Bragança, Escola Superior Agrária, Apartado 1172,
5301-855 Bragança, Portugal
e-mail: mzecast@ipb.pt

N. Evelpidou et al. (eds.), *Natural Heritage from East to West*,
DOI 10.1007/978-3-642-01577-9_18,

"Mirandesa" – a good example of traditional systems and active promotion of products: specimens of this autochtonous cattle breed at contest in Vinhais (photo by F. de Sousa)

Until a few decades ago, the PNM economy was based on agriculture, cattle rearing and several less important activities. Most of the population engaged in traditional stock farming involving few animals. The largely subsistence-based household economy was boosted with income from the sale of animals, eggs, butter and handicrafts. Other important economic activities were smuggling and forestry in Montesinho (Pardo-de-Santayana et al., 2007).

In contrast with many other Mediterranean mountain situations, the modern PNM landscapes are still living; their complex farming systems simultaneously support a multitude of functions other than agricultural production, such as support for recreation, amenity, cultural identity, and preservation of natural resources and environmental quality. The modern landscapes still include the mosaic of meadows, forests, rivers and high mountain vegetation growing on varied geological materials and soils and the predominant natural vegetation consists of oak forest species, broom scrubland and heath. Many fields once used to grow cereals (for bread), pulses, turnip and potatoes now provide grazing for cattle. Agriculture plays only a minor role and new economic activities, such as rural tourism, are increasingly important (IPB/ICN, 2006).

The current and general process of abandonment seems to favour natural ecosystems, promoting the development of habitats of pristine character such as the oak, riparian and chestnuts woods. Field cultivation is not expanding, and planting exotic forest species is not acceptable under current nature conservation regulations. On the other hand burning and grazing are essential process to maintain semi-natural habitats such as scrublands of heather and gorse, as well as the meadows. However burning and grazing must be done in the traditional way, at the right frequency, extension and intensity. If not, they will favour erosion, degradation processes and consequent soil loss. Illegal sand and gravel extraction in rivers and localized foci of water pollution are not frequent but their occurrence prejudices riparian ecosystems.

The great challenge for the near future will be to manage and balance human pressure on natural and semi-natural habitats. On the one hand landscape must be recognized as a dynamic reality that man can only adjust but not rebuild. On the other hand, human processes, such as those involved in PNM landscape dynamics, are associated with rural livelihood and ways of life that cannot be controlled but only be recognized and understood.

Shrubs – persistent elements in PNM landscapes, sign and challenge on land use sustainability: *Cystus ladanifer* near Aveleda, with Bragança urban area and Nogueira Mountain in farther plans (photo by T. de Figueiredo)

An unfortunate example of the difficulties involved in managing these kinds of systems is the system of indemnities or special hunting permitted to farmers that are affected by wildlife incursions into their crops. A few failures of this system have generated hostility against the Natural Park administration. Farmers also use traditional crop-protection methods, which can be improved. For example, some farmers hang large empty cans and a stick from trees. The wind moves the device, which acts as a simple noise-producing mechanism, keeping away wild boars and large wild herbivores. Some farmers put naphthalene in old cans, the smell of which protects vegetable gardens. Electric fences do not exist in the region but they could also be used, although seasonally rather than permanently, and also at the right time, so as to allow people and wildlife to move freely; this is essential to ecological processes (Alves, 2004).

In contrast, a very good example, to be regarded as best practice in rural development, is the promotion of local and protected rural products. In 1980, PNM staff started to organize yearly, in the first weekend of February, the Vinhais Fair of Smoked Sausage, which still continues. Local sausages contain chopped pork meat, seasoned with aromatic herbs or spices (pepper, red pepper, paprika, garlic, rosemary, thyme, cloves, ginger, nutmeg, etc.). Names like "Salpicão", "Alheira", "Linguiça", "Chouriço Verde", "Chouriça Boche", "Chouriço Doce", "Chouriço

Pão", "Chouriço Chaviano", "Butelo", "Presunto", "Chispe", "Orelha" make up a very well developed list of highly profitable products. Currently, the principal agents of commercialization and promotion have been the Town Council of Vinhais and The National Association of Breeders of "Bísaro" pig. These bodies are now in charge of the organization of the Vinhais Fair, without any special need for PNM staff aid (IPB/ICN, 2006).

Vinhais Fair of Smoked Sausage – rural development in practice: exhibition of thewide variety of high quality traditional products in early February (photo by C.Alves, CM Vinhais)

Photos By

F. de Sousa, IPB / ESA, op324539j@mirandesa.pt
C. Alves, Câmara Municipal de Vinhais, carla.alves@cm-vinhais.pt

References

Alves, J. 2004. Man and wild boar: a study in Montesinho Natural Park, Portugal. In Galemys: Boletín informativo de la Sociedad Española para la conservación y estudio de los mamíferos 16(1): 223–230.

Pardo de Santayana, M., Tardio, J., Blanco, E., Carvalho, A.M., Lastra, J.J., San Miguel, E. & Morales, R. 2007. Traditional knowledge of wild edible plants used in the northwest of the Iberian Peninsula (Spain and Portugal): a comparative study. In Journal of Ethnobiology and Ethnomedicine 3(27): 1–11.

IPB/ICN. 2006. Plano de Ordenamento do Parque Natural de Montesinho: I – Relatório de Caracterização. Bragança: Instituto Politécnico de Bragança.

Alvão-Marão: Preserving Important Natural Resources in a Mountain Range

Domingos Lopes, João Bento, Marco Magalhães, and Pedro Ferreira

Introduction

The Alvão-Marão cordillera links the Natura 2000 Network and includes the mountains of Alvão and Marão. These are part of a system of mountains with a north-south orientation, namely Larouco, Alvão, Marão and Meadas. The Alvão-Marão mountain range is bordered to the west by the Tâmega River and to the east by the Corgo River. Its importance has increased since it was integrated into the Natura 2000 Network, although the Portuguese government had previously noted the importance of Alvão-Marão in terms of biodiversity, as will be reported later.

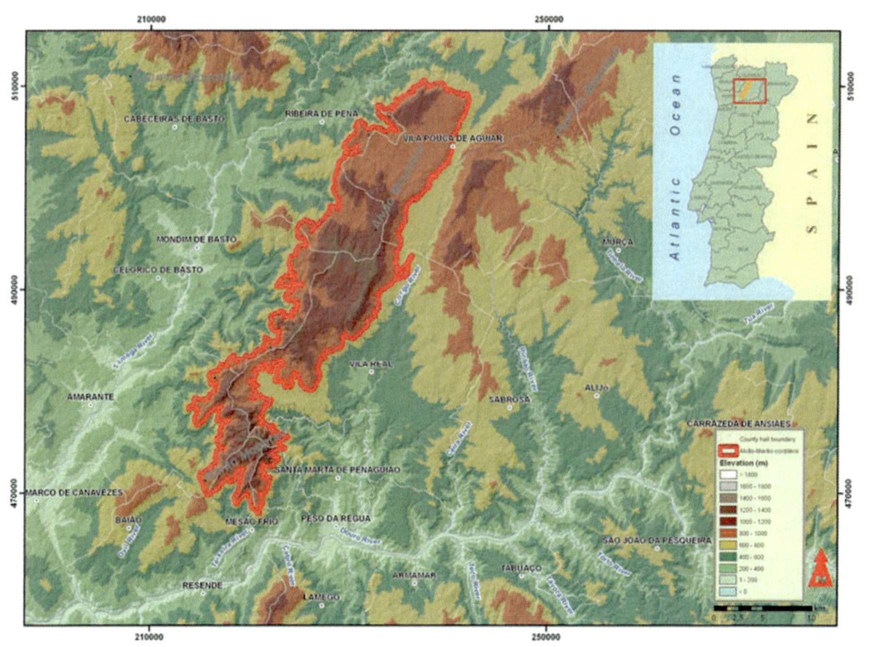

Alvão and Marão: location and altimetry (photo by P. Ferreira)

D. Lopes (✉)
Universidade de Trás-os-Montes e Alto Douro, Departamento Florestal, Vila Real, Portugal
e-mail: dlopes@utad.pt

N. Evelpidou et al. (eds.), *Natural Heritage from East to West*,
DOI 10.1007/978-3-642-01577-9_19, © Springer-Verlag Berlin Heidelberg 2010

Marão is the sixth highest mountain in Portugal, with a maximum altitude of 1,415 m, while Alvão has a maximum elevation of 1,283 m. This cordillera is located in the transition between the "Douro Litoral" (Coastal Douro region) and the "Alto Douro" (Interior Douro region), surrounding Vila Real.

This massive system of mountains is mainly responsible for the continental and extreme climate in all Trás-os-Montes. It works as a barrier to the movement of clouds and air masses from the Portuguese coast inland. It is also a physical barrier for human movements. For many years, before the construction of IP4, the present main road crossing this region, these mountains were a real obstruction, preventing easy access to Trás-os-Montes and Alto Douro. Because of that, the name of this Portuguese region is connected with these two mountains as "Trás-os-Montes" means "behind the mountains" (i.e. Alvão and Marão).

Physical Characterization

Geologically, Alvão and Marão are composed mostly of large bodies of schist and granite. The granitic areas are predominant, and have a strong and direct impact on soils in this region. This type of soil is highly susceptible to erosion and weathering and has low organic carbon.

In terms of climate, the transition from oceanic climate in the Portuguese Minho Region to the Meseta-type climate of the Trás-os-Montes Region occurs in a short distance, across the mountains that separate these two Portuguese provinces. In the latter region the terrain is rough and the winters are cold except along the Douro River, in the hilly vineyard district, where Port wine is produced, and in the warm tributary valleys, where olives and other crops are grown.

There are three main units in the landscape succession: the mountain, predominantly granitic; a transition area between the mountain and the valley, predominantly quartzitic; and the valleys, with lower altitudes but with steep slopes.

Notes on Alvão-Marão Natural Conservation History

The region is and has been protected for its natural values by three institutional instruments, partly overlapping in Alvão-Marão. As mentioned above, it integrates into the Natura 2000 Network. Previously, the territory included forest management areas (called Forest Perimeters by the Portuguese Forest Service), and also a Natural Park in Alvão area, as part of the Portuguese Network of Protected Areas.

A Portuguese law from 1903, that established principles for national forestry, imposed constraints on the land use, preserving biodiversity and justifying this forestry for reasons of public utility. This way, an ecological perspective was already present at this early stage; this partly explains the current ecological importance of this area. The 4th article of that law stated that afforestation could occur not only for commercial timber production, but also to improve water infiltration and the water regime, and also to preserve the hill tops and unused land, which could benefit from the modified climate and reduced soil erosion.

The total area of this mountain range is at least 58,790 ha, distributed over 10 Municipalities. 33,680 ha are covered by forest perimeters (Table 1), established as a result of the implementation of the above mentioned law. In this way, the history of the Portuguese afforestation and the Portuguese Forestry Service is intimately connected with the history of this area.

Table 1 Area of the main Forest Perimeter of Alvão-Marão mountain range and date of the afforestation projects

Forest Perimeter	Total area (ha)	Date of forestation project
Alvão	4,700	1944 (S); 1945 (I)
Serras do Marão (Vila Real) e Ordem	8,816	1956 (S+I)
Serras do Marão e Meia Via	6,400	1916/1939 (S); 1939 (I)
Serra de São Tomé do Castelo	670	1951 (S); 1953 (I)
Mondim de Basto	8,500	1933, 1940 (S); 1939 (I)
Ribeira de Pena	4,500	1944 (S); 1945 (I)

S – Project legal submission; I – Project implementation (source: Germano 2004)

The Natural Park of Alvão was created in 1983, covering a total area of 7,200 ha of which 2,940 and 4,260 ha are, respectively, in the Municipalities of Mondim de Basto and Vila Real.

In terms of a historical perspective of legislation relating to nature protection in Portugal, the first law directly and explicitly focused on nature conservation was published in 1970 and the creation of the first National Park of Peneda-Gerês happened 1 year later. In 1993 Portuguese law extended to conservation activities not only by public but also by private organizations. The need to submit management plans for protected areas was established.

Alvão-Marão has 18 natural habitats, 4 of which have recognized priority for the conservation of nature. The emblematic presence of the wolf (*Canis lupus*) is confirmed and at least 4 packs have been identified in this area (Pimenta et al., 2005).

Natural Resources

Fauna

Until recently, and according to the Portuguese Institute for Nature Conservation (ICN 2006), more than 200 species were surveyed in this region. Approximately 58% of the species are included in Annex II of the Bern Convention, 22% on the list of endangered species of the Vertebrates Red Book of Portugal, and 5% are considered endemic to the Iberian Peninsula.

An incomplete list of some of the most important ones, in terms of protection, is: *Canis lupus* (Iberian Wolf), *Myotis mystacinus* (Whiskered Bat), *Myotis nattereri* (Natterer's Bat), *Tadarida teniotis* (European Free-tailed Bat), *Circaetus gallicus* (Short-toed Eagle), *Falco peregrinus* (Peregrine Falcon), *Anthus spinoletta* (Water Pipit), *Monticola saxatilis* (Rock Thrush), *Ficedula hipoleuca*, *Pyrrhocorax pyrrhocorax* (Red-billed Chough), *Pyrrhula pyrrhula* (Eurasian Bullfinch), *Accipiter gentiles*, *Accipiter nisus* (Eurasian Sparrowhawk), *Strix aluco* (Tawny Owl), *Anthene noctua*, *Buteo buteo* (Common Buzzard), the *Fringila coelebs*, *Regulus ignicapillus* (Firecrest), *Turdus viscivorus* (Mistle Thrush). There are also *Falco tinnunculus* (Common Kestrel), *Bubo bubo* (Eurasian Eagle Owl), *Monticola solitarius* (Blue Rock Thrush) and *Corvus corax* (Common Raven).

Some of the fauna from the Alvão-Marão Cordillera

Flora and Vegetation

The Alvão-Marão cordillera is located in a transition area between the Mediterranean and west European continental regions, a place where two influences merge: that of the wet Portuguese coast and the much drier continental interior. Adding to these effects, altitude introduces one additional influence, which has direct impact on the vegetation covering the region. Thus, the diversity and differentiation of vegetation in these areas is remarkable, with the survey of about 486 species of plants, 25 of them endemic to the Iberian Peninsula, 23 with status of conservation, according to the Natural National Parks Service (ICN 2006).

Over the years, at least until the mid twentieth century, an increasing deterioration of the condition of natural vegetation occurred, which has resulted in an increase in areas of rock outcrops and visible erosion effects (Bento and Fernandes, 1987; Bento et al., 1988; Timóteo et al., 2004).

Land uses for this Mountain range in 1990 and 2000 are shown in Figures 3–4. From analysis of it in recent years, this study area tends to show stable land use because changes were not drastic. The only perceptible change was a slightly increase of shrubs, mainly explained by summer fires (Timóteo et al., 2004).

Also from this analysis it can be confirmed that the *Quercus* stands are extremely important ecosystems for this area, from a botanical and biodiversity point of view. In the areas with Atlantic influence, below 600 m elevation, *Quercus robur* tends to be associated with *Castanea sativa*. In the south facing slopes, species more adapted to Mediterranean climate tend to dominate, like the *Quercus suber*, *Arbutus unedo*, *Laurus nobilis* or *Phyllirea angustifolia*. With the increase of altitude and Continental influences, mixed forests of *Quercus robur* and *Quercus pyrenaica* can be found. The *Betula alba* and *Betula celtiberica* are also very important species, reflecting the altitudinal influence, and some dispersed but representative stands can also be found in this region.

The wet grasslands (called Lameiros in Portuguese), located on the gaps between *Quercus* areas, are ecosystems of extreme importance, because they present a rich and diverse set of herbaceous species.

The preservation of traditional agricultural practices, including the grazing of Lameiros by the maronês cattle (an autochthonous breed whose name is linked to the area of Marão), provides some balance in the plant succession. Maronesa (feminine of maronês) is a breed with a huge potential, because it is well adapted to the regional climate, producing tasty meat and providing an important source of income for local people. Generally, it is important to support these traditional forms of exploitation by the rural communities who live in this area.

The *Quercus* stands in Alvão-Marão (photo by P. Ferreira)

Table 2 Classes of Land use in Alvão-Marão in 1990 and 2000 (Land Cover Corine database)

Types of land cover	Areas in 1990		Areas in 2000	
	(ha)	(%)	(ha)	(%)
Discontinuous urban fabric	0.00	0.00	4.28	0.01
Mineral extraction sites	25.37	0.08	25.37	0.08
Non-irrigated arable land	2065.84	6.24	2065.84	6.24
Permanently irrigated land	133.46	0.40	133.46	0.40
Pastures ("Lameiros")	495.31	1.50	495.31	1.50
Annual crops associated with permanent crops	18.13	0.05	18.13	0.05
Complex cultivation patterns	13.02	0.04	8.74	0.03
Land mainly occupied by agriculture, with significant areas of natural vegetation	1515.82	4.58	1515.82	4.58
Broad-leaved forest	208.59	0.63	226.75	0.69
Coniferous forest	1872.95	5.66	1971.63	5.96
Mixed forest	1960.46	5.93	1692.34	5.12
Natural grassland	2883.01	8.71	2748.10	8.31
Moors and heathland	8295.02	25.07	7780.00	23.52
Sclerophyllous vegetation	384.17	1.16	384.17	1.16
Transitional woodland/shrub	5037.29	15.23	5903.97	17.85
Bare rock	1992.55	6.02	1983.58	6.00
Sparsely vegetated areas	6157.95	18.61	6092.50	18.42
Water bodies	25.09	0.08	34.05	0.10
TOTAL	33084.03	100.00	33084.03	100.00

The Landscape

The landscape is one of the key resources of this region, where high quality eco-tourism can be developed. In this region, wide horizons and beautiful natural landscapes can be found, and houses and villages were built with local stone – e.g. Lamas D'Olo, Ermelo, Barreiro. These small, typical villages are outstanding examples of appropriate human intervention in natural areas.

Alvão-Marão during winter (photo by P. Ferreira)

Agricultural lands in Alvão-Marão (photo by P. Ferreira)

Local stone is also used for sustaining the Fervença terraces, by which farming fits into the orography.

Fields of rye, corn and potatoes emerge in small gaps between the wide rock outcrops which cover extensive areas of these mountains. Lameiros, where the maronês cattle are fed, and communal forested areas, where goats can graze, are also important economic supplements for this rural population, which may disappear in the near future.

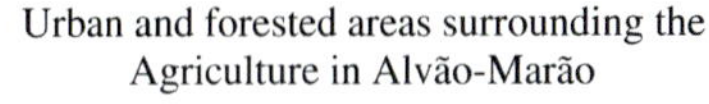

Urban and forested areas surrounding the Agriculture in Alvão-Marão

The Mountain range of Alvão-Marão

Ecotourism is viable in this region, taking advantage of the beauty of the landscape and of the traditions of the people, with great impact on its development (PRODER 2008). Indeed, by definition ecotourism is about connecting conservation, communities, and sustainable travel (TIES 2008). According to the same source, this means that those who implement and participate in responsible tourism activities should follow ecotourism principles:

- minimize impact
- build environmental and cultural awareness and respect
- provide positive experiences for both visitors and hosts
- provide direct financial benefits for conservation
- provide financial benefits and empowerment for local people
- raise sensitivity to the host country's political, environmental, and social climate

References

Bento, J. S. & Fernandes, A. 1987. Estratégias de desenvolvimento local na área do Parque Natural do Alvão – Visão retrospectiva com recurso a matrizes de transição. In *Investigação Operacional* 7(2): 45–54.

Bento, J. S., Fernandes, A. M. & Vicente, A. M. 1988. Formas de ocupação do território no Parque Natural do Alvão (1947–1984). In *Simpósio A floresta e o Ordenamento do Espaço de Montanha*: 243–252. Vila Real: Universidade de Trás-os-Montes e Alto Douro.

Germano, M. A. 2004. Regime Florestal. Um século de existência. In *Estudos e informação* 324. Lisboa: Direcção Geral Recursos Florestais.

ICN. 2006. *Plano Sectorial da Rede Natura 2000*. Lisboa: Instituto de Conservação da Natureza (www.icnb.pt).

Pimenta, V., Barroso, I., Álvares, F., Correia, J., Ferrão da Costa, G., Moreira, L., Nascimento, J., Petrucci-Fonseca, F., Roque, S. & Santos, E. 2005. *Situação Populacional do Lobo em Portugal: resultados do Censo Nacional 2002/2003 (Relatório Técnico)*. Lisboa: Instituto da Conservação da Natureza/ Grupo Lobo.

PRODER. 2008. *Plano Desenvolvimento Rural*. Lisboa: Ministério da Agricultura, Desenvolvimento Rural e Pescas (www.min-agricultura.pt).

TIES. 2008. The International Ecotourism Society. Available at:http://www.ecotourism.org/ (accessed: 20.07.08).

Timóteo, I., Bento, J., Rego, F. & Fernandes A. 2004. Changes in Landscape Structure of the Natural Park of Alvão (Portugal). In S. Mazzoleni et al. (ed.) *Recent Dynamics of the Mediterranean Vegetation and Landscape*: 211–216. London: John Wiley and Sons, Ltd.

A Groundwater System in a Mountain Environment (Serra da Estrela, Portugal)

Jorge Espinha Marques, José Manuel Marques, and Calos Aguiar

Mountains are often recognized as the "world's water towers", due to their exceptional ability to generate water resources of good quality and, frequently, of considerable economic value (e.g. UNESCO IHP-VI Programme: http://www.unesco.org, Aureli 2002).

Such areas are the source of most of the larger river systems all over the world, and usually represent some of the blackest "black boxes" in the hydrological cycle. The seasonality and spatial variability of local groundwaters and the complex role of geomorphology, geology, climate, soils, land use and human activities on the hydrology of mountain areas are rather difficult to model, even when relevant data are available. Nevertheless, mountain river basins provide the best opportunity to increase knowledge on the relationship between those complex variables as well as their impacts on the water quality in zones at different elevations, under different cultural settings (Chalise, 1994).

Serra da Estrela is the highest mountain in the Portuguese mainland (with an altitude reaching 1993 m a.s.l.) and is part of the Cordilheira Central, an ENE-WSW mountain range that crosses the Iberian Peninsula. This region shows specific climatic, geologic and geomorphologic features that play an important role on the local water cycle, where surface waters, normal groundwaters and thermomineral waters occur side-by-side.

In order to understand the hydrogeological system of the river Zêzere Basin Upstream of Manteigas village (ZBUM), an area of particular interest in this mountain, an integrated multidisciplinary approach has been taken, comprising geology, geomorphology, climatology, hydrogeology, geochemistry, isotope hydrology, hydropedology and geophysics, under the scope of the HIMOCATCH R&D Project "Role of High Mountain Areas in Catchment Water Resources, Central Portugal" (e.g. Espinha Marques et al., 2006a, 2007).

The catchment under study corresponds to an area of ca. 28 km^2 with an altitude ranging from 875 m a.s.l., at the stream flow gauge measurement weir of Manteigas,

J.E. Marques (✉)
Universidade do Porto, Faculdade de Ciências, Departamento de Geologia/Centro de Geologia, Porto, Portugal
e-mail: jespinha@fc.up.pt

N. Evelpidou et al. (eds.), *Natural Heritage from East to West*,
DOI 10.1007/978-3-642-01577-9_20, © Springer-Verlag Berlin Heidelberg 2010

to 1993 m a.s.l., at the Torre summit. The relief of the study region consists mainly of two major plateaus, separated by the NNE-SSW valley of the Zêzere river (Vieira, 2008). Late Pleistocene glacial landforms and deposits are a distinctive feature of the upper Zêzere catchment, since the majority of the plateau area was glaciated during the Last Glacial Maximum (e.g. Daveau et al., 1997; Vieira, 2008). The Serra da Estrela climate has Mediterranean features, with mean annual precipitation reaching 2,500 mm in the most elevated areas (Daveau et al., 1997). Precipitation seems to be mainly controlled by the slope orientation and the altitude (Mora, 2006). Mean annual air temperatures are below 7°C in most of the plateau area and, in the Torre vicinity, they may be as low as 4°C.

The main regional hydrogeological units are: (i) sedimentary cover, including alluvium and quaternary glacial deposits; (ii) metasedimentary rocks, which include schists and graywackes; and (iii) granitic rocks. In the ZBUM area the main lithotype is granite, followed by sedimentary and metasedimentary rocks, resulting in water circulation media that are dominantly fractured, rather than porous. Locally, in the alluvium and quaternary glacial deposits as well as in the most weathered granites and metasedimentary rocks, the porous media are dominant. Porous media usually occur at shallower depths (typically less than 50 m). The main regional tectonic structure is the Bragança-Vilariça-Manteigas Fault Zone (BVMFZ).

The study of the unsaturated zone was carried out from a hydropedologic perspective (e.g. Espinha Marques et al., 2007). The pedologic units that occur in this sector are: (i) Humic, Leptic and Skeletic Umbrisols; (ii) Lithic and Umbric Leptosols; (iii) Umbric Fluvissols; (iv) rock outcrops. The spatial distribution of the soil features depends, mainly, on the parent material, relief, climate and land cover. In the ZBUM, the prevailing soil profile is ACR type, with an Umbric A horizon. These soils are coarse textured, acid, with very high A horizon organic matter content (essentially controlled by altitude); its total porosity values are greater than the reference ones for coarse soils. Water retention in the A horizon is strongly controlled by the organic matter content. Field saturated hydraulic conductivity in the A horizon is high and the behaviour of unsaturated hydraulic conductivity is typical of coarse soils. In spite of the high permeability of the A horizon, soils with high runoff potential prevail.

Mathematical hydrological modeling has proven to be an essential tool to understand the ZBUM hydrogeology. This task was carried out by means of the hydrological package Visual Balan v2.0 (e.g. Espinha Marques et al., 2006b). This achieved the following distribution of the mean yearly precipitation (of 2,336 mm) in the hydrological years between 1986–87 and 1994–95: 41.1% corresponding to interflow, 14.9% to groundwater discharge, 13.0% to overland flow and 31.0% to current evapotranspiration and interception. Thus, the aquifer recharge is estimated to be around 15% of the mean yearly precipitation. The high value of interflow is typical of montane hydrogeological systems.

The hydrogeochemical study identified normal groundwaters, thermomineral and surface waters (e.g. Espinha Marques et al., 2006a; Marques et al., 2007). Normal groundwaters and surface waters are hypothermal, hyposaline, with pH between 5 and 7 and sodium-bicarbonate or sodium-chloride facies (which, in the

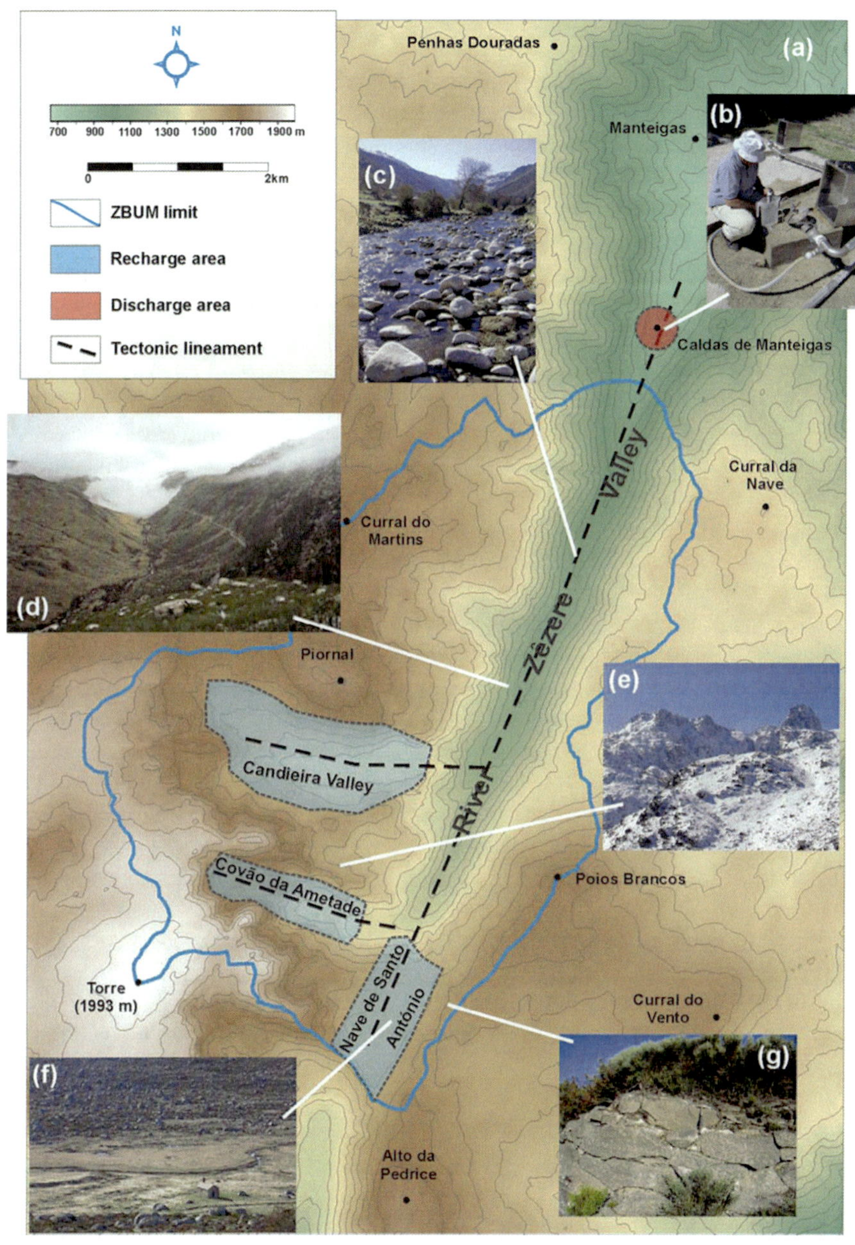

(a) Hypsometric features of the Zêzere Basin Upstream of Manteigas, showing the recharge and discharge areas of the thermomineral system; (b) thermomineral water sampling; (c) Zêzere river; (d) Zêzere river U-shaped valley; (e) snow covered granitic peaks; (f) alluvium and glacial deposits; (g) soil profile (Leptosol)

latter case, could result from pollution by salt, spread over roads to allow fast snow melt). Thermomineral waters are sulphide rich, hyperthermal (with maximum measured temperature of 48°C), hyposaline (yet, more mineralized than normal groundwaters), with pH around 9 and sodium-bicarbonate facies. Geothermometry provided an estimate of the maximum depth of thermomineral water circulation of around 3.1 km. The Local Meteoric Water Line was determined as well as a $\delta^{18}O$ gradient of –0.142‰ /100 m of altitude (which indicates that the recharge of the thermomineral subsystem takes place between 1,400 and 1,600 m a.s.l.). The absence of H^3 and the C^{14}-age dating pointed out that the infiltration of normal groundwaters is undoubtedly more recent than the infiltration of thermomineral waters.

The conceptual hydrogeological model proposed for the ZBUM encompasses the whole local hydrologic cycle and identifies three interrelated types of aquifers: shallow aquifers (unconfined and with normal groundwater circulation), intermediate aquifers (semi-confined to confined – with normal groundwater circulation) and deep aquifers (confined – with thermomineral water circulation). The thermomineral aquifer recharge results from the downward water circulation through permeable zones associated with the BVMFZ and conjugate structures. Three areas where this recharge possibly takes place were identified, all situated between 1,400 and 1,600 m a.s.l.: the Nave de Santo António area, the Covão da Ametade and the Candieira valleys. The thermomineral subsystem features suggest that the fluid circulation is deep and slow, taking about 10,000 years to travel from the recharge areas to the discharge at Caldas de Manteigas.

This study has highlighted the Serra da Estrela mountain region as a source of strategic groundwater resources, connected to a complex hydrogeological system. The ZBUM area, in particular, supplies high quality drinking water for a major bottling industrial unit and for domestic use at Manteigas village. The presence of important thermomineral water resources made it possible to create and exploit the Caldas de Manteigas Spa. Furthermore, the groundwater discharge to the river Zêzere flow contributes to the storage at the Castelo do Bode great dam, the main source of Lisbon's water supply.

Therefore, the hydrogeological study carried out in Serra da Estrela helps to support the idea that mountain groundwater systems are essential for sustainable development due to their social, economic and environmental influence.

References

Aureli, A. 2002. What's ahead in UNESCO's International Hydrological Programme (IHP VI 2002–2007)?. In Hydrogeology Journal 10: 349–350.

Chalise, S. R. 1994. High mountain hydrology in changing climates: perspectives from the Hindu Kush-Himalayas. In L. Molnár, P. Miklánek & I. Mészáros (eds.) Developments in hydrology of mountain areas. Proceedings FRIEND AMHY Annual Report No.4, IHP-V, Technical Documents in Hydrology 8: 23–31. New York: UNESCO.

Daveau, S., Ferreira, A. B., Ferreira, N. & Vieira, G. 1997. Novas observações sobre a glaciação da Serra da Estrela. Estudos do Quaternário (Lisboa) 1: 41–51.

Espinha Marques, J., Duarte, J. M., Constantino, A. T., Martins, A. A., Aguiar, C., Rocha, F. T., Inácio, M., Marques, J. M., Chaminé, H. I., Teixeira, J., Samper, J., Borges, F. S. & Carvalho, J. M. 2007. Vadose zone characterisation of a hydrogeologic system in a mountain region: Serra da Estrela case study (Central Portugal). In L. Chery & Gh. de Marsily (eds.) Aquifer Systems Management: Darcy's Legacy in a World of Impending Water Shortage (Selected papers on Hydrogeology, IAH, Volume 10): 207–221. London: Taylor & Francis Group.

Espinha Marques, J., Marques, J. M., Chaminé, H. I., Carreira, P. M., Fonseca, P. E., Samper, J., Vieira, G. T., Mora, C., Teixeira, J., Martins Carvalho, J., Sodré Borges, F. & Rocha, F. T. 2006a. Hydrogeochemical model of a low temperature geothermal system in a mountainous terrain, Serra da Estrela, Central Portugal. In Geothermal Resources Council Transactions 30: 913–918.

Espinha Marques, J., Samper, J., Pisani, B. V., Alvares, D., Vieira, G. T., Mora, C., Martins Carvalho, J., Chaminé, H. I., Marques, J. M. & Sodré Borges, F. 2006b. Avaliação de recursos hídricos através de modelação hidrológica: aplicação do programa Visual Balan v2.0 a uma bacia hidrográfica na Serra da Estrela (Centro de Portugal). In Cad. Labor. xeol. Laxe (A Coruña) 31: 86–106.

Marques, J.M., Carreira, P. M., Espinha Marques, J., Chaminé, H. I., Fonseca, P. E., Monteiro Santos, F., Almeida, E., Gonçalves, R., Pierszalik, R., Martins Carvalho, J., Almeida, P. G., Cavaleiro, V. & Teixeira, J. 2007. A multitechnique approach to the dynamics of thermal waters ascribed to a granitic hard-rock environment (Serra da Estrela, Central Portugal). In H. Marszalek, K. Chudy (eds.) Selected Hydrogeologic Problems of the Bohemian Massif and of other Hard Rock Terrains in Europe, Acta Universitatis Wratislaviensis, seria: Hydrogeologia No 3041. Wroclaw: Wydawnictwo Uniwersytetu Wroclawskiego.

Mora, C. 2006. Climas da Serra da Estrela, características regionais e particularidades locais dos Planaltos e do Alto Vale do Zêzere. PhD Thesis. Lisbon: University of Lisbon. 427 pp.

Vieira, G. 2008. Combined numerical and geomorphological reconstruction of the Serra da Estrela plateau icefield, Portugal. Geomorphology 97: 190–207.

Northwest Beira Highlands – Freita and Caramulo Hills (Portugal)

Celeste Oliveira Alves Coelho, Sandra Valente, and Cristina Ribeiro

Introduction

To the South of the Douro and within the Vouga river systems, not far from the Atlantic coast, the Northwest Beira Highlands rise steeply from the coastlands and plains. They are composed of small mountains, but have a very marked relief because the Hercynian Massif (Maciço Antigo), after suffering repeated folding, granitization and metamorphism, has been greatly worn down by erosion (Ferreira, 1978). They include the Montemuro Mountain in the South Douro river system, the Gralheira Massif (Freita, Arada and Arestal hills), and the Caramulo Mountain to the South (see location map of Portuguese continental highlands in the introductory article to this chapter).

In 30–40 km, from the coast, the relief rises rapidly from almost sea-level to over 1,000 m, intercepting the moist oceanic air masses and producing abundant rainfall (1,500–2,000 mm) in autumn, winter and spring. A hot and dry summer season is present from June to September; this is characteristic of transitional Atlantic Mediterranean climates. Winter months, in the highlands, have several days with temperatures below freezing (Coelho, 2006).

The contrast between the landscape of granitic mountains with deep incised valleys, in the western part of the Beira region, and the undulating plateaus inland, to the east, also with a marked Atlantic influence, is also visible in dominant vegetation types.

The natural vegetation cover of high ground has been worn down by centuries of overgrazing and fire, and at present the more representative species are: *Erica cinerea* L., *Erica umbellata* L. and *Calluna vulgaris, Erica arborea* L. and *Chamespartium tridentatum* L. In less disturbed sites, *Genista florida* L. and *Cytisus striatus, Genista triacanthus* Brot. and *Ulex minor* Roth occur. Small patches of oaks (*Quercus robur* L. and *Quercus pyrenaica* Wild are the remnants of the Atlantic

C.O.A. Coelho (✉)
Department of Environment and Planning, CESAM – Centre for Environmental and Marine Studies, University of Aveiro, Portugal
e-mail: coelho@dao.ua.pt

N. Evelpidou et al. (eds.), *Natural Heritage from East to West*,
DOI 10.1007/978-3-642-01577-9_21,

dominant forests. Along the riverside *Alnus glutinosa* (L) Gaertner., *Salix* spp. and *Fraxinus angustifolia* Vahl. would be present, but today they remain in only the more inaccessible places (Moura, 2001).

The majority of the natural forest cover has been substituted by *Pinus pinaster* Aiton, and in the recent past by *Eucalyptus globulus* Labill. plantations, which have become the more abundant species in the Northwest Mountains of Beira, in particular in the Caramulo Mountain.

The prevailing soil types are Cambisols and Leptosols and in general are considered of limited suitability for agricultural uses.

Arqueological evidence of human occupation of the highlands demonstrates the importance of these areas since Neolithic times. Early settlers lived on the higher ground. The present distribution of population shows the marked influence of relief: under 400 m it is dispersed, but from that level up to 800 m the settlement is concentrated in small hamlets and villages (Moura, 2001). The population has been decreasing since the middle of the twentieth century, nowadays there is no substitution by the younger generation and the population is ageing. Many small hamlets and villages have been abandoned. The main activities are subsistence farming of winter cereals, irrigated maize, fodder and pasture. In the communal lands (baldios) grazing by small ruminants was the main economic activity of the mountains, in particular in the higher plateaus. The afforestation of baldios, by the State, in the early years of last century, reduced the area available for pasture for traditional herds of small ruminants, and increased wild fire risk.

Serra Da Freita Geopark

The Serra da Freita is part of the Gralheira massif, oriented Northwest to Southeast, with an area of circa 350 km^2 in the Arouca, Vale de Cambra and São Pedro do Sul municipalities. Freita stands out from the Precambrian schists very strongly.

Freita has certain geologic and cultural features which due to their singularity and scenic landscape value are the basis for the creation of a geopark to be approved by UNESCO. The Geoparque Arouca has the following aims: (i) the study and conservation of the geological heritage of the region; (ii) the promotion and valorization of the natural and cultural heritage for the local communities and the general public; (iii) the raising of awareness of schools and the younger generation of the conservation of natural heritage; (iv) the promotion of eco-tourism and the valorization of traditional goods and services; and (v) the establishment of partnerships between the different stakeholders (Sá et al., 2007; GTGA, 2006, http://www.cm-arouca.pt).

Geomorphological and geological features associated with high landscape quality and traditional activities are relevant in Freita, such as:

- the transition from the granite to micaschist rocks is, in many areas, very marked by erosion as in the Mizarela falls (a vertical waterfall of 60 m, in the Caima river);

The Mizarela falls (Freita)

- the granite outcrops of Castanheira, with nodules surrounded by biotite, with a biconvex disc shape of up 15 cm, looking like eggs. Locally they are called deliverance rocks (pedras parideiras) since they are detached from the main rock, due to weathering. They are very rare; only one other similar geological site is known, in Russia;
- the presence of unique trilobites fossils of the Middle Ordovician in Pedreira do Valério;
- the high landscape diversity associated with the steep slopes, agricultural terraces, and the undulating plateaus, with its associated flora and fauna;
- the agricultural fields and the grazing grounds of Arouquesa (a local beef-cattle breed) raised on the fertile soils of the lower slopes of the mountain.

Landscape diversity in the Serra da Freita

Caramulo Mountain: Traditional Water Systems

Caramulo Mountain is a narrow and long relief feature of circa 500 km^2, representing well the landscape associated with granite geomorphological features. The more marked characteristics of the Caramulo relief are the very steep slopes facing east. Caramulo Mountain is included in the Águeda, Oliveira do Hospital, Vouzela and Tondela municipalities.

Landscape diversity in the Caramulo Mountain

In the past, the main activities in Caramulo were agro-sylvo-pastoral systems. Nowadays, the population still depends on subsistence farming and forestry. Water resources are abundant in the granite massif, due to the rainfall pattern, over 2,000 mm per year, and to the presence of large volume groundwater storage associated with granite weathering. Local communities have developed, since ancient times, water harvesting, storage and distribution systems in order to improve irrigated crops and to provide domestic water supply. These systems use the geological and geomorphological setting of the granite landscape (Valente et al., 2008). Some of the more representative systems are:

- springs, usually associated with a small system of channels ("levadas") that flow by gravity and distribute water to the terraces and storage tanks;
- small embankments and dikes, which store and convey water until they can discharge enough to irrigate the maize fields in summer and to protect grasslands from frost in winter ("água de lima");

"Levadas" in Caramulo

- fountains used for domestic supply;
- pits for groundwater harvesting;
- São João do Monte aqueduct in granite, which is fed by natural spring and used to transport water to the fields.

Traditional water systems have played an important role in the sustainability of the area and will be very relevant in the future, especially when the rainfall pattern is modified by climate change.

Fountains in Caramulo

Concluding Remarks

The Freita and Caramulo hills in the Central Region represent an important natural and cultural heritage associated with landscape and traditional land uses typical of Atlantic Mediterranean transitional climates. However, these areas, in particular the Serra da Freita, are suffering an enormous pressure from neighbouring urban visitors, which endangers the survival of fauna, flora and geological sites, calling for emergency action to control the new demands in these areas. The recognition of the Freita Geopark by UNESCO could represent an opportunity to reverse the depletion of natural resources (Valente and Figueiredo, 2003).

The abandonment of the subsistence agro-sylvo-pastoral systems, together with ageing and depopulation, jeopardizes the sustainability of traditional water systems. Nevertheless, the modernization of these systems could be important for the maintenance of the cultural landscape of these areas.

References

Coelho, C. 2006. Soil Erosion in Portugal. In J. Bordman & J. Poesen (eds.) Soil Erosion in Europe: 359–367.London: Wiley.

Ferreira, A. B. 1978. Planaltos e Montanhas do Norte da Beira. Mem. Cent. Est. Geog. n. 4. Lisboa: Centro de Estudos Geográficos.

Guia de Portugal 1984–1985. I – Beira Litoral, II – Beira Baixa e Beira Alta, 2nd ed.. Lisboa: Fundação Calouste Gulbenkian.

Moura, A. R. 2001. A Serra da Freita. Aveiro¨129.

Ribeiro, O. 1954. Portugal: O Mediterrâneo e o Atlântico. Lisboa: Sá da Costa.

GTGA 2006. Projecto para a candidatura do Geoparque Arouca à Rede Europeia de Geoparques. Arouca: Grupo de trabalho do Geoparque Arouca. Available at: http://www.cm-arouca.pt (accessed: 30/01/2006).

Sá, A. A., Brilha, J., Cachão, M., Couto, H., Medina, J., Rocha, D., Valério, M., Rábano, I. & Gutiérrez-Marco, J. C. 2007. Arouca Geopark: a new project towards the sustainable development based on the conservation and promotion of the geological heritage. In VII Congresso de Geologia: 893–896. Évora: Universidade de Évora.

Valente, S. & Figueiredo, E. 2003. O turismo que existe não é aquele que se quer. . . In O. Simões & A. Cristóvão (org.) TERN – Turismo em Espaços Rurais e Naturais: 95–106. Coimbra: Instituto Politécnico de Coimbra.

Valente, S., Coelho, C., Ferreira, A., Amaral, L. & Silva, F. 2008. Traditional water management systems – Are they still important to local communities? In Geophysical Research Abstracts Vol. 10. Vienna: EGU General Assembly 2008.

Ecological and Cultural Consequences of Agricultural Abandonment in the Peneda-Gerês National Park (Portugal)

Yvonne Cerqueira, Cláudia Araújo, Joana Vicente, Henrique Miguel Pereira, and João Honrado

Agricultural Abandonment in Marginal Farmlands – Drivers and Consequences

The socioeconomic and technological development of human societies has been driving profound changes in the classic relationships between cities and the countryside (Gutman, 2007). In the context of such changes, the abandonment of agricultural land is a growing concern throughout much of Europe's rural mountain areas. Migration and aging population are the main reasons behind the collapse of traditional farming systems and the increase in land abandonment. Together with an intrinsic resistance to adopting modern market-oriented farming practices, these processes induce consequences (still poorly evaluated) to the environment as well as numerous socio-economic impacts. Many of the most traditional types of agricultural landscapes in Europe are dramatically decreasing due to this partial or complete abandonment of farmland (EEA, 2005). Some of the most critical conservation issues today relate to changes in traditional farming practices on habitats such as hay meadows, lowland wet grasslands, dry grasslands and arable land. Overall, these habitats usually disappear after the abandonment of traditional farming practices; as a result species adapted to the diversity of structures or resources in such High Nature Value (HNV) farmlands may not survive (Henle et al., 2008). Consequently these simplified, homogenous landscapes resulting from agricultural abandonment usually have greatly reduced biodiversity when compared to the more diverse HNV farmland areas.

The preservation of traditional biotopes and landscapes such as pasturelands is desirable and significant for biodiversity. The strong international concern regarding the consequences of agricultural abandonment results from the recognition that the most important anthropogenic cause of agro-ecosystem biodiversity loss is rapid change in land use and land cover, together with the subsequent transformation of

Y. Cerqueira (✉)
CIBIO – Centro de Investigação em Biodiversidade e Recursos Genéticos, Departamento de Botânica, Faculdade de Ciências da Universidade do Porto, Portugal
e-mail: yvonne.cerqueira@fc.up.pt

N. Evelpidou et al. (eds.), *Natural Heritage from East to West*,
DOI 10.1007/978-3-642-01577-9_22,

habitats (MEA, 2005; Perrings et al., 2006). The magnitude of the impacts of such modifications is expected to be influenced by other types of environmental shifts e.g. climatic changes, which are forecasted to act synergistically with other promoters of land use change (Abildtrup et al., 2006; Schumacher and Bugmann, 2006). Even though some of the most valuable elements of terrestrial biodiversity (e.g. forest core biodiversity) may be favoured by such changes, it is assumed that agricultural abandonment has a globally negative impact on biodiversity because it naturally leads to more homogeneous vegetation cover. This process, known as vegetation succession, usually results in a structural change from an open to a closed landscape, causing: the loss of small-scale mosaics of diversified land use and of their characteristic species. There is also a loss of species related to forest edge (ecotone) habitats, a reduction in genetic diversity in both wild species and in local breeds of livestock or varieties of crops, and an increased fire risk in the landscape context, since abandoned grazing areas no longer act as firebreaks (CAP, 2004). Agricultural abandonment, producing such homogeneous landscapes, is also expected to affect the diversity and level of ecosystem services, although ecosystem/landscape properties such as resistance to alien invasion (which is forecasted to be promoted in case of climatic warming; Chytry et al., 2008) are still poorly evaluated.

According to the IRENA report (EEA 2005), many recommendations relative to soil cover, crop rotation, cultivation practices and management of crop residues have been included in the regulations of Portugal. However, there are also indications of a strong flow from agriculture to forest land cover classes (EEA, 2005). Abandonment has affected many types of farmland including significant areas of high natural value. The abandonment of agricultural land has been driven by a variety of pressures linked to economic and social transition. Protected areas are part of the core strategy to conserve biodiversity, but many studies demonstrate that reserves alone will not be sufficient to conserve all biodiversity. Conservation strategies will be critical in places used for commodity production such as multiple-use forests (Lindenmayer and Franklin, 2002) and agricultural areas where the management of different components of vegetation cover within areas broadly designated for agriculture can make a significant contribution to the persistence of native species from many *taxa* (Sirami et al., 2007). For that reason, one important feature of agriculture landscapes is the role of resident local farming or pastoral communities as key stewards, decision makers and managers of biodiversity (Scherr and McNeely, 2008).

Agricultural Abandonment and Landscape Changes in the Peneda-Gerês National Park

The Peneda-Gerês National Park (PNPG), in the northwest of Portugal, is an area of high natural value with a diverse array of flora, fauna, ecosystem types, and landscapes (Honrado, 2003). It is a protected area in which traditional agricultural practices were evident but are now declining due to the abandonment of marginal farmlands or to increasing specialization or intensification on lowland, more fertile, agricultural land. In Portugal, this phenomenon started in the 1960s when many immigrated to other European countries in search of a better life. Then in the 1980s many returned and agricultural activities increased once again. It wasn't until the early 1990s that many of the young farmers abandoned the practice due to lack of government support and adaptive capacity, and it was during this time that the aging of rural population also accelerated (MADP, 2003).

When looking at the socio-economic structure of the Park in the late 20th century, the primary sector employed a total of 52% of the Park residents, the tertiary sector held a total of 24.6%, and finally the secondary sector with 22.9%. In the 1980s and early 1990s, the primary sector suffered a decrease of 15% in its activity, with an opposite trend recorded for the secondary and tertiary sectors (PNPG, 1995). However, even though agricultural activities suffered a strong decrease, they still remain the dominant activity. Unfortunately marginal agriculture does not generate enough income for a comfortable lifestyle and very little investment is attributable to this activity. As a result most of the younger and older generations have left these landscapes in search of a more comfortable lifestyle which in turn has weakened the already fragile social-economic structure, characterised by an aging population dominated by women. In the primary sector, activities related to cattle raising (breeding) have the best conditions since subsidies are granted (PNPG, 1995). However, activities connected to tourism, hydroelectric power and forestry now generate the most capital. The tertiary sector is the second most important sector in the Park (PNPG, 1995), with coffee shops, restaurants and hotels found throughout the Park generating revenue and job opportunities.

The cultural landscapes of the Peneda-Gerês National Park have been shaped through centuries of traditional agricultural practices and comprise many endemic elements of biodiversity (Honrado, 2003). However, recent changes in land use, mostly related to agricultural land abandonment, may well be threatening the capacity of these landscapes to host biodiversity in a sustainable condition (Moreira et al., 2001; Metzger et al., 2006; Reidsma et al., 2006). At the same time, they are also increasing fire susceptibility (Moreira et al., 2001) and promoting agricultural intensification or afforestation of these systems. In the past, extensive marginal lands were cleared in order for herds and cattle to graze and pass through, and wood was gathered for human use. Due to the new demographic structure these areas are now covered with dense brush increasing the occurrence of fires. This increase affects biological communities in that it leads to strong modifications in the once complex vegetation mosaics. This in turn has consequences to the fauna, which is dependent on these locations for food and shelter, and may also promote soil erosion and degradation. Agricultural land abandonment also leads to the degradation or disappearance of traditional agricultural infrastructures e.g. terraces ("socalcos") and traditional irrigation systems ("regadios"). These traditional infrastructures are important for soil preservation and if ignored they will collapse and the system will lose its functionality. Moreover, the loss of such infrastructures leads to the disappearance of these valuable cultural landscapes. Other typical traditional activities in this area, such as herding and cattle raising, are also in decline. In the 1950s there were in the Park roughly 60,000 cattle of the "barrosã" breed, typical of the Northern mountains of Portugal, but currently there are only about 8,000 animals (MADP, 2003).

Another relevant but sometimes forgotten consequence of agricultural land abandonment is the loss of Traditional Ecological Knowledge (TEK), which has been the primary source of information used by local communities throughout centuries to manage the land in a way that sustained biodiversity and controlled the overall ecology of landscapes. At present time the rate of erosion of TEK and of traditional agricultural knowledge in rural communities of the National Park is unknown and so its consequences to the conservation of biodiversity and on the provision of ecosystem services are still poorly known. It is therefore imperative that this knowledge possessed by the elderly people be collected and integrated with modern ecological knowledge as this synthesis may provide the answer to long-term sustainability of these areas of high priority conservation value. Although rural land abandonment should be viewed as something negative, it nevertheless has positive aspects. For example, it can present potential benefits for biodiversity, providing the opportunity to correct or rectify the degradation of habitats provoked by the intensification of agriculture in the last several decades. The abandonment of agricultural practices results in the encroachment of natural vegetation. In the National Park this regeneration is mainly dominated by oak forests, tall scrublands, and riparian woods. A decrease in cattle herding also signifies a decrease in pressure on natural habitats and wildlife populations. On the other hand, some habitats and priority species listed in the EU "Habitats" and "Birds" Directives may be threatened by such encroachment processes e.g. those associated with open, deforestated landscapes.

For centuries, rural mountain inhabitants have developed agricultural and silvi-pastoral techniques which permitted the exploitation of natural resources in an apparently sustainable way. However, modern socio-economic and political drivers, which have been the force behind changes in land use, are inducing shifts in the properties of ecosystems and landscapes, which in turn control the conditions and trends of ecosystem services (Pereira et al., 2004, 2005; Metzger et al., 2006). In fact, habitats normally associated with agricultural practices provide various functions, goods and services that are linked to human well-being, even if public perception of such services and other aspects of landscapes is complex and often contradictory (Nijnik et al., 2008). A study carried out in Sistelo, a rural community on the border of the National Park, as a part of the Millennium Ecosystem Assessment (Pereira et al., 2004), identified the connection of human well-being and ecosystem services at the local level. The study provided researchers with an outline of 40 different criteria for human well-being. It also revealed that independency of local provisioning services was an improvement in quality of life in which many no longer were required to work the land in order to obtain goods and services (Pereira et al., 2005). In Sistelo, locals viewed the landscape (which holds cultural and aesthetic value at the national scale) and cultivated terraces ("socalcos", as they are known) in the mountain slopes, as something negative. They also stated that levelling these landscapes would improve human well-being. On the other hand they identified traditional cultural practices as having an important role in developing social capital and enhancing social well-being (Pereira et al.,

2004), since those practices were responsible for bringing together the local people to work in one another's fields. But they also agreed that material well-being improved mainly due to access to new income sources such as retirement pensions. However, although there may be an improvement in general well-being, the sustainable supply of ecosystem services in the future may well decline and so quantifying and mapping the current provision of such services is crucial to support decision making and to guide future land management (Egoh et al., 2008).

Future Perspectives For Tek, Biodiversity and Ecosystem Services in Rural Landscapes of the Peneda-Gerês National Park

Low intensity farming, which is the case for most European mountain areas, is viewed as beneficial for the preservation of biodiversity, and so maintaining these agro-ecosystems is of high importance (Bignal and McCracken, 1996; Moreira et al., 2001; Metzger et al., 2006; Reidsma et al., 2006). In Europe, many species depend on semi-natural landscapes, which have been created and maintained by long-term low intensity land use. Today, changes in agriculture cause a loss of species-rich ecosystems that depend on traditional land use (Poschlod et al., 2005). Particularly on High Nature Value (HNV) farmland, abandonment involves significant losses of biodiversity, because its characteristic species strongly depend on low inputs of fertilizers and on grazing or mowing (CAP, 2004). In marginal landscapes, abandonment leads to a loss of diverse, open landscapes due to secondary succession following the absence of anthropogenic disturbance, and to a decrease of landscape complexity and compartmentalization (Pereira et al., 2004). Different management regimes can be applied to control secondary succession and thus minimize habitat and biodiversity loss (García, 1992; Muller, 2002). In practice, managers have to compromise between ecological benefits and financial costs of management schemes.

There is a growing consensus that functional diversity is important to short-term ecosystem resource dynamics and long-term ecosystem stability, as it increases positive interactions or complementary functions (Diaz and Cabido, 2001; Lavorel et al., 2008). European low intensity farming, in the form of livestock rearing and traditional cultivation methods, has created semi-natural habitats that now support a range of species, such as species-rich grasslands, hay meadows, and grazed wetlands. Many of the habitats that are valued now for biodiversity are a direct result of traditional agricultural practices established during agricultural expansion. High Nature Value (HNV) farming systems and their associated management practices can be highly beneficial to biodiversity (Bignal and McCracken 2000; EEA, 2004), and the maintenance of functional diversity in many ecosystems depends directly on traditional types of agricultural land use and farming practices, which therefore contribute to the maintenance of ecosystem services on a local scale. It is important to ensure that the intensity of agricultural management is appropriate for biodiversity conservation (Bignal and McCracken, 2000), and integrated, systematic assessments are required for such complex realities as agroecosystems (Moonen and Bàrberi 2008).

Agricultural land abandonment is a problem that can no longer be ignored and one which will affect a significant percentage of mountain areas such as the Peneda-Gerês National Park. It is imperative that in these regions of high risk political measures be taken to reduce the effects of land abandonment. The most serious conflict in Portugal is the divergence of conversion of extensive farmlands into high intensity production areas and the loss of large areas of such extensively used agricultural landscapes through abandonment (Moreira et al., 2001). Current land use changes have prompted land abandonment in areas where agriculture was considered uneconomical, and negatively affect biodiversity. Abandonment is widespread in several regions of Portugal, leading to a lessening of disturbance caused by humans and favouring natural vegetation dynamics. The closing of the parts of the landscape that were maintained open by grazing or agriculture and the cessation of traditional wood extraction for firewood or charcoal are expected to cause a progressive reduction in contrast among the patches that defined the earlier landscape mosaic. Currently land abandonment has also initiated a simplification of the traditional landscape mosaic. If this trend towards a simplification of the landscape is to be mitigated, the challenge for managers will be to find ways to maintain mosaics in which open habitats, local heterogeneity and woodlands co-exist (Ross et al., 2008). This should increase the diversity of birds and other animals and help maintain biodiversity in the face of land abandonment.

Conservation management is strongly constrained by economic considerations. The major pressures affecting farmland biodiversity have often been agricultural support policies, but the current changes to the CAP support mechanism are expected to indirectly result in a decrease in pressure on biodiversity. The Agenda 2000 reform introduced the opportunity for farmers to obtain support (under the Rural Development Plan) for additional activities other than farming *per se*, provided they comply with a range of EU directives (including the Birds and Habitats directives) and maintain their land in good agricultural and environmental condition. Farmsteads are thus expected not only to produce food but also to promote biodiverse and culturally rich landscapes. The modern concept of ecoagriculture further recognizes that agriculture-dependent rural communities are critical stewards of biodiversity and ecosystem services. While protected natural areas are essential in ecoagriculture landscapes, to ensure critical habitat for vulnerable species, maintain water sources and provide cultural resource, these resources often may be owned or managed by local communities and farmers. Current trends suggest that a continuing and growing demand for agricultural and wild products and ecosystem services will require farmers, agricultural planners and conservationists to reconsider the relationship between crop production and conservation of biodiversity (Scherr and McNeely, 2008). In fact, a number of EU policies support the management of abandoned land for the benefit of biodiversity. The CAP comprises two principal forms of support – direct payments available to nearly all farmers, and a range of selective payments for rural development measures. The continued management of HNV farmlands, and the restoration of some fairly limited areas of abandoned land, is a major environmental priority. It is important that the funds provide opportunities to address the issue of agricultural land abandonment, not only through direct support to restoration and biodiversity sensitive management, but also for agricultural land to be continuously managed in a way that preserves, and possibly rises, its biodiversity.

Many of the recent changes in the socio-economic structure of the Peneda-Gerês National Park have been primarily related to the rapid increase in agricultural land abandonment, and they are thought to induce losses of both traditional knowledge

Aspects of rural landscapes and of their biodiversity in the Peneda-Gerês National Park (PGNP). Top row: The traditional, culturally rich rural landscapes of the PGNP are dominated by semi-natural hay meadows and native oak forests. Middle row: Agricultural and pastoral abandonment leads to forest encroachment in both agricultural lands (left) and extensive pasturelands (right). Bottom row: Abandonment of herding, cattle raising and agriculture may lead to the disappearance of valuable elements of local biodiversity such as mires (left) and endemic plant species living in forest edges and low intensity meadows (*e.g. Paradisea lusitanica*, right)

and biodiversity in these mountain landscapes. Political drivers have also played an important part in this phenomenon and immediate action is necessary as human well-being depends on the sustainability of ecosystem services provided by these landscapes. Even though there is still a large amount of uncertainty concerning the responses of ecosystems and their services to ongoing environmental changes (particularly concerning climatic shifts, land use change, and their synergetic effects with other drivers; Schröter et al., 2005; Metzger et al., 2006), conservation perspectives are fairly positive in the Peneda-Gerês National Park, since an ongoing revision of the Park master plan, based on a systemic view of natural and cultural landscapes, is affirmatively putting ecosystem services at the core of land management and conservation planning. This may well provide the backbone of a sustainable opportunity to balance nature conservation with human well-being through the proper valuation of social structures and traditional knowledge.

References

Abildtrup, J., Audsley, E., Fekete-Farkas, M., Giupponi, C., Gylling, M., Rosato, P. & Rounsevell, M. 2006. Socio-economic scenario development for the assessment of climate change impacts on agricultural land use: a pairwise comparison approach. In *Environmental Science and Policy* 9: 101–115.

Bignal, E.M. & McCracken, D.I. 1996. Low-intensity farming systems in the conservation of the countryside. In *Journal of Applied Ecology* 33: 413–424.

Bignal, E.M. & McCracken, D.I. 2000. The nature conservation value of European traditional farming systems. In *Enviromental Reviews* 8: 149–171.

CAP 2004. *Common Agricultural Policy – Seminar on Land Abandonment and Biodiversity, in relation to the 1st and 2nd pillars of the EU's Common Agricultural Policy*. Sigulda, Latvia, 7–8 October 2004.

Chytry, M., Maskell, L.C., Pino, J., Pysek, P., Vilà, M., Font, X. & Smart, S.M. 2008. Habitat invasions by alien plants: a quantitative comparison among Mediterranean, subcontinental and oceanic regions of Europe. In *Journal of Applied Ecology* 45: 448–458.

Díaz, S. & Cabido, M. 2001. Vive la différence: plant functional diversity matters to ecosystem processes. In *Trends in Ecology and Evolution* 16(11): 646–655.

EEA, European Environment Agency 2004. *High nature value farmland: characteristics, trends and policy challenges*. Report N 1/2004, Copenhagen.

EEA, European Environment Agency 2005. *Agriculture and environment in EU-15—the IRENA indicator report*. EEA Report, no. 6/2005, Copenhagen.

Egoh, B., Reyers, B., Rouget, M., Richardson, D.M., Le Maitre, D.C. & van Jaarsveld, A.S. 2008. Mapping ecosystem services for planning and management. In *Agriculture, Ecosystems and Environment* 127: 135–140.

García, A. 1992. Conserving the species-rich meadows of Europe. In *Agriculture, Ecosystems and Environment* 40: 219–232.

Gutman, P. 2007. Ecosystem services: Foundations for a new rural–urban compact. In *Ecological Economics* 383–387.

Henle, K., Alard, D., Clitherow, J., Cobb, P., Firbank, L., Kull, T., McCrackeng, D., Moritz, R.F.A., Niemelä, J., Rebane, M., Wascher, D., Watt, A. & Young, J. 2008. Identifying and managing the conflicts between agriculture and biodiversity conservation in Europe – A review. In *Agriculture, Ecosystems and Environment* 124: 60–71.

Honrado, J. 2003. Flora e vegetação do Parque Nacional da Peneda-Gerês. Ph.D. thesis, Universidade do Porto.

Lavorel, S., Grigulis, K., McIntyre, S., Williams, N.S.G., Garden, D., Dorrough, J., Berman, S., Quetier, F., Thebault, A. & Bonis, A. 2008. Assessing functional diversity in the field – methodology matters! In *Functional Ecology* 22(1): 134–147.

Lindenmayer, D.B. & Franklin, J.F. 2002. *Conserving forest biodiversity: a comprehensive multiscaled approach.* Washington, D.C.: Island Press.

MADP, Ministério da Agricultura, Desenvolvimento Rural e Pescas, 2003. *O Abandono da Actividade Agrícola.* Grupo de Trabalho Agro-Ambiental. Lisboa.

Metzger, M.J., Rounsevell, M.D.A., Acosta-Michlik, L., Leemans, R., Schröter, D. 2006. The vulnerability of ecosystem services to land use change. In *Agriculture, Ecosystems and Environment* 114:69–85.

Millennium Ecosystem Assessment (MA) 2005. *Ecosystems and human well-being: synthesis.* Washington, DC: World Resources Institute.

Moonen, A.-C., & Bàrberi, P. 2008. Functional biodiversity: An agroecosystem approach. In *Agriculture, Ecosystems and Environment* 127: 7–21.

Moreira, F., Rego, F.C., Ferreira, P.G. 2001. Temporal (1958–1995) pattern of change in a cultural landscape of northwestern Portugal: implications of fire occurrence. In *Landscape Ecology* 16: 557–567.

Muller, S. 2002. Diversity of management practices required to ensure conservation of rare and locally threatened plant species in grasslands: a case study at a regional scale (Lorraine, France). In *Biodiversity and Conservation* 11: 1173–1184.

Nijnik, M., Zahvoyskab, L., Nijnik, A. & Oded, A. 2008. Public evaluation of landscape content and change: Several examples from Europe. In *Land Use Policy* 26: 77–86.

Pereira, H.M., Domingos, T. & Vicente, L. eds. 2004. *Portugal Millennium Ecosystem Assessment: State of the assessment report.* Centro de Biologia Ambiental, Faculdade de Ciências da Universidade de Lisboa.

Pereira, E., Queiroz, C., Pereira, H.M., & Vicente, L. 2005. Ecosystem services and human well-being: A participatory study in a mountain community in Northern Portugal. In *Ecology and Society* 10(2): 14.

PNPG 1995. *Plano de Ordenamento do Parque Nacional Peneda-Gerês (Relatório de Síntese).* Braga.

Poschlod, P., Bakker, J.P. & Kahmen, S. 2005. Changing land use and its impact on biodiversity. In *Basic and Applied Ecology* 6: 93–98.

Reidsma, P., Tekelenburg, T., van den Berg, M. & Alkemade, R. 2006. Impacts of land-use change on biodiversity: An assessment of agricultural biodiversity in the European Union. In *Agriculture, Ecosystems and Environment* 114: 86–102.

Ross, B., Cunningham, D., Lindenmayer, B., Crane, M., Michael, D., MacGregor, C., Montague-Drake, R. & Fisher, J. 2008. The Combined Effects of Remnant Vegetation and Tree Planting on Farmland. In *Bird Conservation Biology* 22(3): 742–752.

Scherr, S.J. & McNeely, J.A. 2008. Biodiversity conservation and agricultural sustainability towards a new paradigm of 'ecoagriculture' landscapes. In *Philosophical Transactions of the Royal Society B* 363: 477–494.

Schröter, D., Cramer, W., Leemans, R., Prentice, I.C., Araújo, M.B., Arnell, N.W., Bondeau, A., Bugmann, H., Carter, T.R., Gracia, C.A., de la Vega-Leinert, A.C., Erhard, M., Ewert, F., Glendining, M., House, J.I., Kankaanpää, S., Klein, R.J.T., Lavorel, S., Lindner, M., Metzger, M.J., Meyer, J., Mitchell, T.D., Reginster, I., Rounsevell, M., Sabaté, S., Sitch, S., Smith, B., Smith, J., Smith, P., Sykes, M.T., Thonicke, K., Thuiller, W., Tuck, G., Zaehle S. & Zierl, B. 2005. Ecosystem Service Supply and Vulnerability to Global Change in Europe. In *Science* 310:1333–1337.

Schumacher, S. & Bugmann, H. 2006. The relative importance of climatic effects, wildfires and management for future forest landscape dynamics in the Swiss Alps. In *Global Change Biology* 12:1435–1450.

Sirami, C., Brotons, L. & Martin, J.-L. 2007. Vegetation and songbird response to land abandonment: from landscape to census plot. In *Diversity and Distributions* 13: 42–52.

The Cyclops Islands

Laura Maria Padovani

On the eastern coast of Sicily, below the Aetna volcano, there is a tiny archipelago, a marine protected area, where the legend of Ulysses is still alive

"ἔνθεν δὲ προτέρω πλέομεν ἀκαχήμενοι ἦτορ: Κυκλώπων δ' ἐς γαῖαν ὑπερφιάλων ἀθεμίστων ἱκόμεθ', οἵ ῥα θεοῖσι πεποιθότες ἀθανάτοισιν οὔτε φυτεύουσιν χερσὶν φυτὸν οὔτ' ἀρόωσιν, ἀλλὰ τά γ' ἄσπαρτα καὶ ἀνήροτα πάντα φύονται, πυροὶ καὶ κριθαὶ ἠδ' ἄμπελοι, αἵ τε φέρουσιν οἶνον ἐριστάφυλον, καί σφιν Διὸς ὄμβρος ἀέξει. τοῖσιν δ' οὔτ' ἀγοραὶ βουληφόροι οὔτε θέμιστες, ἀλλ' οἵ γ' ὑψηλῶν ὀρέων ναίουσι κάρηνα ἐν σπέσσι γλαφυροῖσι, θεμιστεύει δὲ ἕκαστος παίδων ἠδ' ἀλόχων, οὐδ' ἀλλήλων ἀλέγουσιν."

"νῆσος ἔπειτα λάχεια παρὲκλιμένος τετάνυσται, γαίης Κυκλώπων οὔτε σχεδὸν οὔτ' ἀποτηλοῦ, ὑλήεσσ': ἐν δ' αἶγες ἀπειρέσιαι γεγάασιν ἄγριαι: οὐ μὲν γὰρ πάτος ἀνθρώπων ἀπερύκει, οὐδέ μιν εἰσοιχνεῦσι κυνηγέται, οἵ τε καθ' ὕλην ἄλγεα πάσχουσιν κορυφὰς ὀρέων ἐφέποντες."

"ὣς ἐφάμην, ὁ δ' ἔπειτα χολώσατο κηρόθι μᾶλλον, ἧκε δ' ἀπορρήξας κορυφὴν ὄρεος μεγάλοιο, κὰδ δ' ἔβαλε προπάροιθε νεὸς κυανοπρῴροιο. ἐκλύσθη δὲ θάλασσα κατερχομένης ὑπὸ πέτρης: τὴν δ' αἶψ' ἤπειρόνδε παλιρρόθιον φέρε κῦμα, πλημυρὶς ἐκ πόντοιο, θέμωσε δὲ χέρσον ἱκέσθαι. αὐτὰρ ἐγὼ χείρεσσι λαβὼν περιμήκεα κοντὸν ὦσα παρέξ, ἑτάροισι δ' ἐποτρύνας ἐκέλευσα ἐμβαλέειν κώπης, ἵν' ὑπὲκ κακότητα φύγοιμεν, κρατὶ κατανεύων: οἱ δὲ προπεσόντες ἔρεσσον."

Thence we sailed on with aching hearts, and came to the land of the Cyclops, a rude and lawless folk, who, trusting to the immortal gods, plant with their hands no plant, nor ever plough, but all things spring unsown and without ploughing, wheat, barley, and grape-vines with wine in their heavy clusters, for rain from Zeus makes the grape grow. Among this people no assemblies meet; they have no stable laws. They live on the tops of lofty hills in hollow caves; each gives the law to his own wife and children, and for each other they have little care.

Now a rough island stretches along outside the harbor, not close to the Cyclops' coast nor yet far out, covered with trees. On it innumerable wild goats breed; no tread of man disturbs them; none comes here to follow hounds, to toil through woods

L.M. Padovani (✉)
Department of Biotechnology, Agro-industry and Health Protection, ENEA – National Agency for New Technology, Energy and the Environment Casaccia Research Center, 00123 Rome, Italy
e-mail: padovani@casaccia.enea.it

N. Evelpidou et al. (eds.), *Natural Heritage from East to West*,
DOI 10.1007/978-3-642-01577-9_23,

and climb the crests of hills. The island is not held for flocks or tillage, but all unsown, untilled, it evermore is bare of men and feeds the bleating goats...

So I spoke, and he was angered in his heart the more; and tearing off the top of a high hill, he flung it at us. It fell before the dark-bowed ship a little space, but failed to reach the rudder's tip. The sea surged underneath the stone as it came down, and swiftly toward the land the wash of water swept us, like a flood-tide from the deep, and forced us back to shore. I seized a setting-pole and shoved the vessel off; then inspiriting my men, I bade them fall to their oars that we might flee from danger, with my head making signs, and bending forward, on they rowed.

The Cyclops Islands include Lachea, the Faraglione Grande, the Faraglione Piccolo and the Faraglione degli Uccelli, formed of basaltic lava of the Etna (Aetna) volcano, characterized by crystallized columns with peculiar shapes on which limestone incrustations and marine erosion have produced astonishing effects. Lachea, the only true island, is just in front of the entrance to the port of Aci Trezza, only 200 m away, on the Sicilian coastal tract known as the Cyclops Riviera.

History

The tiny archipelago of the Cyclops Islands (Isole dei Ciclopi or Isole Lachee) has always been evoked by a mythological scene of Homeric tradition. The legend identifies the islets as the stones flung by the Cyclops Polyphemus against Odysseus (Ulysses) and his companions on the vessel.

A painting of the episode described in Homer, Odyssey, Book IX

The archipelago bears testimony of human occupation that goes back to prehistory. The archaeological importance of these places has been famous since the end of the 1800s after the discovery ceramic material of the prehistoric, protohistoric, Roman and Byzantine periods, and of Medieval grottoes and anthropogenic chambers. Also recent investigations (1999 and 2000), carried out in collaboration with the Cultural Heritage Department of Catania, have led to the discovery of cockpits dug into the cliff, of uncertain use, and ceramic materials of late Roman date.

Traces of prehistoric structures (holes for poles dug into the rocks) have also been found. Materials have been collected underwater, especially in the bay of Aci Trezza; these are likely evidence of the presence of an ancient anchorage, perhaps favored by the shelter offered by the *faraglioni* to boats during storms.

The archipelago and the Aci Trezza coast as well were the property, since 1671, of the Riggio of Campofiorito princely dynasty; Lachea was given in emphyteusis at the beginning of 1800; in 1869, the owner, senator Luigi Gravina, donated it to the University of Catania, to be used for scientific studies.

The harbor of Aci Trezza with the Isola Lachea (left) and the Faraglioni's (right)

Basic Information

The archipelago and the surrounding water is nowadays a Marine Protected Area along the eastern (Ionian) coast of Sicily in the Province of Catania. The size of the Area is as follows:

	Surface (a)	Coastline (m)
Zone A	35	814
Zone B	202	1,475
Zone C	386	4,240
Total	623	6,529

The Protected Area is for the most part in Acitrezza (Aci Trezza or simply Trezza) within the municipality of Aci Castello and for a small part in Acireale. The name Aci is also linked to the legend of the Cyclops. In Greek mythology, Ace (Akis) was a minor pastoral deity, loved by the Nereid (sea nymph) Galatea, coveted by Polyphemus: the jealous Cyclops killed the rival crushing him with a rock.

The Protected Area was instituted by government decree on December 7th 1989 to protect the extraordinary wealth of sea organisms of the area, as well as its beauty and cultural heritage. The Area is managed by a consortium composed of the C.U.T.G.A.N.A. of the University of Catania and the Municipality of Aci Castello. Since 2001 this consortium, named "Isole dei Ciclopi", has been safeguarding an inestimable natural patrimony and also an archaeological heritage from various periods. The area is subdivided into three zones according to differing degrees of protection: area "C" is a partial reserve area, "B" is a general reserve area, and "A" is a full reserve area.

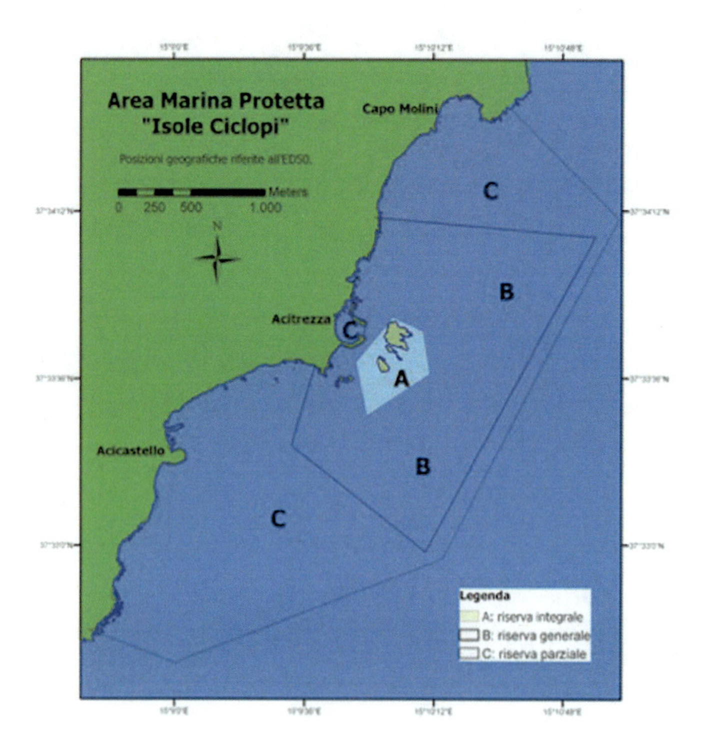

The island of Lachea is part of the small archipelago, also named Lacheo Archipelago or Isole Lachee, facing the coast of Acitrezza. "L'Isola" (the island), as it is commonly called by the local people, has an irregular shape with a surface area of more than 2 ha: the side facing Acitrezza is approximately 250 m long. The higher parts of the island are covered by clays of sandy color resting on the basaltic formation. In the front part, reachable by stone stairs, there is an old building where the fish museum is located, with an old sink and a small dwelling dug into the hardened clay, that probably was the dormitory of Saint John the Anchorite, a hermit of the end of the XI century. At sea level, facing Acitrezza, there is another building that is used by the caretaker.

From a geological point of view, the coast in front of the archipelago is famous for the occurrence of the so-called "pillows". Pillows are volcanites originating from submarine eruptions of magma; in the past, Italian scholars have called them "globular basalts". Today, they are known as pillows or pillows lavas, because of their spherical shape. They can also be observed in Iceland, New Zealand, and the Iblea area of Sicily, but the pillows of Acitrezza have been described as those with the most characteristic and extraordinary shapes. Pillows are formed when magma erupts through a fracture in the sea bed. Seawater cools the emerging magma changing it into a vitreous black crust that is subsequently broken by the pressure of ascending gases, forming various shapes of spheroids.

A similar lava formation, the Hawaiian "Pahoehoe" (where is possible to walk bare-foot) cannot be mistaken for pillows, as it is without vitreous covering and lacking the characteristic radial structure. Pillows can be observed in the hill of Aci Castello and in Acitrezza, north of the old harbour, in two zones where they have a typical conformation. Those in Aci Castello are more compact, while those in Acitrezza have been changed by the furious action of the sea over thousands of years, so that only the roots are still intact.

Columnar basalts, closely connected to the pillows, are also present; these massive basaltic formations indicate the occurrence of now extinct submarine volcanoes. Similar columnar basalts can be admired in the gorges of Alcantara in Sicily, in Antrim in Northern Ireland (the famous "Giants Causeway"), in Fingal's Cave in the island of Staffa in Scotland and, again in Sicily, in Motta Sant'Anastasia, where they form an extraordinary gorge, a relic of an old volcano. They are pentagonal or hexagonal columns and can be disposed vertically (like organ pipes) as in the southern part of Lachea, horizontally as on the Scardamiano seafront in Aci Casello, or in a fan shape, as with those on the central *faraglione*.

Fauna and Flora

Lachea Island is also famous for a rare reptile living on the island, isolated from the mainland of Sicily. The lizard called *Podarcis sicula ciclopica* is an endemic subspecies, that remained for thousands of years separated from the species present on the mainland, and has had a peculiar evolution. There are several types of invertebrates like Collembola (springtails and snow fleas), Coleoptera (beetles),

Hymenoptera (bees and wasps), and Lepidoptera (butterflies). Among the Arachnida (spiders) Gnaphosidae, there is in particular the *Urozelotes mysticus*, the only example on the island. Also interesting, among the birds nesting on the island, are the so-called "*passera sarda*", the Sardinian hen-sparrow. In winter, the island becomes the refuge of seagulls and cormorants.

The flora of the Lachea was certainly damaged by the exploitation of the land for agriculture, since the beginning of 1800, when it was given in emphyteusis. The short distance from the coast has favored colonization by plant species able to tolerate the difficult saline environmental conditions. Today, among the species living on the island, the following can be noted: *Sonchus oleraceus, Ferula communis, Senecio vulgaris, Chrysanthemum coronarium* and the highest *Ailanthus*. A particular species is *Catapondium marinum*, typical of the zones in contact with marine water.

Marine Environment

The sea banks between the islands and the coast are sandy and the water does not exceed 12 m in depth. Numerous large rocks, forming canyons and coves, and in the southern area emerging from the sea, characterize the area. The island and the *faraglioni* (sea cliffs strongly eroded by the sea now isolated from the mainland) emerge from the sandy bottom with sheer walls surrounded by fallen rock blocks that, towards the northern, southern and eastern sides, reach 25 m under the sea.

Different views of the Cyclops Islands landscape

Deep underwater channels characterize the eastern sea banks of Lachea. On the hard substrates (rock walls, masses and pebbles), in the first few metres, many species of algae are present, dominated by the brown alga of the *Cystoseira* kind and by *Astroides* (a kind of star-coral). In deeper waters, near the coral bearing zone, Poriphera are numerous (*Axinella damicornis, Agelas oroides, Spongia officinalis* and the false coral), while less abundant is the yellow Gorgonian. On the debris-covered sea banks, Echinodermata (such as the sea lily and the brittle star) are common. In these areas fish are also present, including White Sea bream, *Dentex*, European bass, grouper, and even sword-fish and tuna - once the nutritional and commercial basic catch of the local fishermen – together with sardines, anchovies and sand lances.

Traditional Products and Craftwork

The area is known for fish and the sea food. Fishing has been the traditional activity and for a long time has been the only means of survival of the local people, with the use of various fishing tools such as:

- *tremaglio* or trammel net (a small-meshed net inserted between two large-meshed nets),
- *cianciolo* (a net dipped into the sea to make a circular shape, close at the bottom to form a semisphere),
- *conzi* (long fishing lines with numerous hooks),
- nets to fish near the rocks,
- *lampara* (fishing boat equipped with lamps to attract fish and cephalopods),
- *polpare* (fishing line with hooks with a white decoy to catch octopus).

Today traditional fishing activity has been abandoned and new types of activities have been substituted, in particular the tourism industry; recreational fishing tours, with fishing boats, called *pesca-turismo*, are significant.

Typical food products are Etna red and white wines, sweets and pastries, the famous granite (grated frozen lemon juice or coffee), *cannoli* and *cassatelle* of ricotta cheese. The claim should also be kept in mind that in the XVII century, following old traditions based on the use of Aetna snow, a fisherman of Acitrezza, Francesco Procopio dei Coltelli, developed modern ice cream, introducing the use of sugar and proposing the utilization of a mixture of salt and ice to keep the product frozen.

Websites

www.acitrezzaonline.it/
www.galenfrysinger.com/acitrezza_sicily.htm
www.parchipertutti.it/public/dl/ExhibitsAMP.rtf
www.santacaterinahotel.com/eng/rivieradeiciclopi.htm
www.volcanoetna.com/en/nature/the-lachea-island-and-faraglioni.html

The Gate of Hades: The Phlegraean Fields

Paola Carrabba

With its volcanic craters and volcanic/lagoon lakes on top of a giant active caldera Averno in ancient times was believed to be the Gate of Hades.

Spelunca alta fuit uastoque immanis hiatus,
scrupea, tuta lacu nigro nemorumque tenebris,
quam super haud ullae poterant impune uolantes
tendere iter pennis: talis sese halitus atris
faucibus effundens supera ad conuexa ferebat
[unde locum Grai dixerunt nomine Aornum].
(Publius Vergilius Maro, Aeneis, Liber VI)

Era un'atra spelonca, la cui bocca
fin dal baratro aperta, ampia vorago
facea di rozza e di scheggiosa roccia.
Da negro lago era difesa intorno,
e da selve ricinta annose e folte.
Uscia de la sua bocca a l'aura un fiato,
anzi una peste, a cui volar di sopra
con la vita agli uccelli era interdetto;
onde da' Greci poi si disse Averno.
(Virgilio, Eneide, Libro VI, traduzione di Annibal Caro)

Deep was the cave; and, downward as it went
From the wide mouth, a rocky rough descent;
And here the' access a gloomy grove defends,
And there the' navigable lake extends,
O'er whose unhappy waters, void of light,
No bird presumes to steer his airy flight;
Such deadly stenches from the depths arise,
And steaming sulphur, that infects the skies.
From hence the Grecian bards their legends make,
And give the name Avernus to the lake
(Virgil, The Aeneid, Book VI, translated by John Dryden)

P. Carrabba (✉)
Department of Biotechnology, Agro-industry and Health Protection, ENEA – National Agency for New Technology, Energy and the Environment, Casaccia Research Center, 00123 Rome, Italy
e-mail: carrabba@casaccia.enea.it

N. Evelpidou et al. (eds.), *Natural Heritage from East to West*,
DOI 10.1007/978-3-642-01577-9_24, © Springer-Verlag Berlin Heidelberg 2010

History

Just to the west of Naples, Italy, there is a volcanic area, a large complex of craters and fumaroles, known as the *Campi Flegrei*, or the Phlegraean Fields. "Burning fields" is the name given by the Greeks to an area extending beyond the present *Campi Flegrei* to Vesuvius. It was where Greek mythology placed the battle between the Giants and the Olympic gods. As no major volcanic eruption is known to have occurred during Greek times (excluding the activity of Ischia), it is likely that the Greeks adapted the legends of the former inhabitants to their own mythology.

The first Greek colonies were founded around the middle of the eighth century BC in Ischia and Cuma (Cumae) by colonists from the cities of Kumai and Chalkis in Euboea (Evia). One of the earliest examples of the classical Greek alphabet has been found in Cuma on the so-called Cup of Nestor. Cuma attained its maximum power during the fifth century BC when, allied with the Greeks of Syracuse, it defeated a coalition of Etruscans and Carthaginians. After the naval battle of Cuma, the Syracusans settled in Ischia, but were soon driven out by an eruption which occurred on the island.

During the IV and III centuries BC, the area fell under the influence of Rome, which, after the Samnite wars, became the ruler of peninsular Italy. During this time, *Campi Flegrei* was a malarial area, covered by marshes and by a thick forest called "Silva Gallinaria", often the refuge of bandits. The area of Baia and Pozzuoli (Puteoli, "little hot springs", the most important port of the Roman Empire) became the resort area of the Roman aristocracy after the destruction of the main part of the Silva Gallinaria to build up the fleet of Augustus. The emperors' palace, where many of the murders of Nero occurred, was situated near the present *Punta dell'Epitaffio*, in Baia, but is now submerged at a depth of 8 m. Another important port was Misenum, where a Roman military navy fleet was stationed (the same that, under the command of Pliny the Elder, carried out a rescue operation at Pompei during the notorious Vesuvius eruption in 79 AD). Little is known of the area in the Middle Ages, when part of the coast was submerged; during this time, the villages suffered from Arab incursions, until they became part of the Kingdom of Naples. Since then, the area of *Campi Flegrei* has followed the vicissitudes of the rest of southern Italy. Most of the district remained a malarial swamp until the beginning of this century.

Geology

The Phlegraean Volcanic District has an extremely complex structure, due to repeated volcanic activity within a limited area. The landscape is distinctive, with numerous extinct craters and continuing volcanic activity, connected with the phenomenon of bradyseism (Greek for "slow earthquake", referring to decades to centuries long gradual uplift or descent of the earth surface) and hot spring. The geological history of the *Campi Flegrei* complex has been subdivided into three periods:

General view of the Campi Flegrei (note the number of craters)

Period I – The oldest period (37,000 years ago). It ended with the eruption of the *Campanian Ignimbrite* which covered approximately 30,000 km^2 with 150 km^3 of magma; this event is suggested to have been the most dramatic event within the Mediterranean area over the past 200,000 years. It caused the collapse of the first caldera, the largest one, covering the present Fields, the southern portion of Naples, the northern part of the Bay of Naples and the entire Pozzuoli Bay.

Period II – Between 37,000 and 14,000 years ago. During this period, at least 11 explosive eruptions took place with the formation of a second caldera. The most important event during this period was the *Neapolitan Yellow Tuff* eruption. It was the second largest eruption in the Campania area, covering an area of approximately 1,000 km^2 with 40 km^3 of magma, ad it also caused the collapse of the caldera.

Period III – This period covers the last 12,000 years. The last eruption occurred in 1538 AD, with the formation of a new cinder volcano, the Monte Nuovo, a cone that grew to 123 m high in 1 week. The caldera floor at this time was on average 60 m lower than at the present, with the larger part of caldera depression being invaded by the sea. Only the northern sector remains constantly above sea level. The majority of the eruptive vents were located within the northeastern sector of the *Neapolitan Yellow Tuff* caldera.

The Phlegraean Fields are therefore characterized by very infrequent, although potentially catastrophic, events. The magmatic chamber is still there, under the sea, evidenced by the fumaroles, tremors and bradyseismic crises.

Part of the caldera is now accessible on foot. It contains a large number of fumaroles, from which steam can be seen issuing, and a number of pools of boiling mud. Several subsidiary cones and tuff craters lie within the caldera.

Serapeum, Pozzuoli

The caldera has a long history of uplift and subsidence. It is, in fact, a classic locality of bradyseismic activity, as recorded in historical documents and by once submerged, then elevated, and now once again submerged Roman structures. The famous *Serapeum*, the Roman marketplace in Pozzuoli, has two generations of floors, the second built when the first one sank below sea level.

Slow uplift and subsidence are also known from other points in the Bay of Pozzuoli. Roman ruins are also found below sea level along the northern and western shores of the Bay, near Monte Nuovo (Baia) and Cape Miseno, respectively.

The Phlegraean Fields' long history of dramatic ground movements is unsurpassed anywhere in the world. Since Roman times, the elevation of the caldera floor has varied by more than 12 m; in the 48 hours before the most recent eruption in 1538 (Monte Nuovo), the floor rose by at least 4–5 m.

Craters and Lakes

In the caldera there are numerous extinct craters, such as *Solfatara*, *Astroni* and *Monte Gauro*. The Romans already knew *Solfatara* in Imperial times. Strabo (66 BC–24 AD) gives the most ancient written testimony coming to us, in his "Strabonis geographica", about the supposed entrance to Hades. It was included among the forty most famous thermae of the Phlegreaen Fields since the Middle Ages. There was no traveller of the eighteenth and nineteenth century who would not include *Solfatara* among his excursions within the so-called "Grand Tour", a form of educational tourism for the young scions of the European noble families.

The *Astroni* volcano was active in the *Campi Flegrei* caldera until 3,800 year ago. Nowadays *Astroni* is a protected area, a green space near the city of Naples.

Monte Gauro is the higher volcanic crater of *Campi Flegrei.* It is well preserved only in northern and northwestern sectors. The southern sector was affected by extractive activities, due to the presence of many tuff caves.

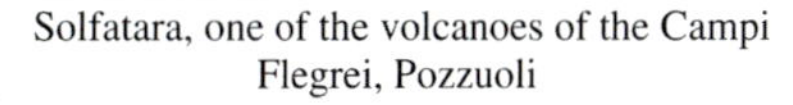

Solfatara, one of the volcanoes of the Campi Flegrei, Pozzuoli

The "Big Lake" in "*Astroni* Natural Reserve"

Campi Flegrei is also characterized by the presence of lakes. One of the craters of the caldera is filled by *Lake Averno* (Avernus). The lake is the crater of a volcano formed 3,700 years ago within an older volcano named *Archiaverno. Averno* was considered by the Greeks and Romans to be the entrance to Hell (the Gate of Hades), because of the abundant fumes rising from it. The name of the lake, in Greek, means "with no birds", as they dropped dead when flying into the gases exhaled from the water.

Lake Averno

Joseph Mallord William Turner, 1814–1815, Äneas und die Cumaeische Sibylle

Monte Nuovo is very close to the lake, toward the sea. The "new mountain" is the cone produced by the last eruption of *Campi Flegrei*, in 1538. In the background, there is the so-called temple of Apollo, which was not destroyed by the low-energy eruption. However, the eruption did cause the destruction of a small village called Tripergole.

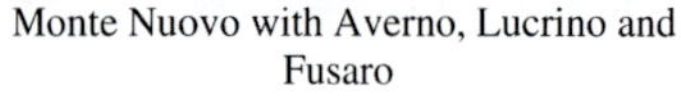
Monte Nuovo with Averno, Lucrino and Fusaro

Lake Fusaro seen from Scalandrone hill

Lake Lucrino is a natural basin near *Lake Averno*. It was the place for a luxury holiday during the Roman period, when many *domus romanae* were built. We can find mention of the lake in the writing of several ancient Roman writers, such as Cicero, Pliny the Elder, Horace, Martial, Propertius, Virgil. The size of the lake was dramatically reduced by the Monte Nuovo eruption in 1538.

Lake Lucrino and Lake Averno

Lake Miseno (also called "Maremorto") is a coastal lagoon formed within a crater, where the Roman fleet sheltered during the winter season. It was thought to be the mythical Stygia Palus (swamp) of Hades. Styx (from which Stygia), Phlegethon, Acheron, and Cocytus were believed to be the four rivers converging at the center of Hades, where two other legendary rivers, Lethe and Eridanos, also ran.

Vanvitelli's royal shooting lodge on Lake Fusaro

Lake Fusaro, a regular coastal lagoon, was considered to be the mythical Acherusia Palus (from Acheron). During the reign of the Bourbon king Ferdinand IV of the Two Sicilies, it was utilized as hunting reserve. Nowadays it is exploited for mussel farming.

References

Strabo. 1923. The Geography of Strabo, 8 vol, The Loeb Classical Library. Cambridge Mass: Harvard University Press, London.

Pliny the Elder. 1952. Natural History, 12 vol, The Loeb Classical Library. Cambridge Mass: Harvard University Press, London.

Silius Italicus. 1934. Punica, 2 vol, The Loeb Classical Library. Cambridge Mass: Harvard University Press, London.

Dionisio di Alicarnasso. 1984. Storia di Roma Arcaica, Rusconi Editore.

Barberi, F., Cassano, E., La Torre, P. and Sbrana, A. 1991. Structural evolution of Campi Flegrei Caldera in light of volcanological and geophysical data. J. Volcanol. Geotherm. Res., 48(1/2): 33–49.

Berrino, G., Corrado, G., Luongo, G. and Toro, B. 1984. Ground deformation and gravity change accompanying the 1982 Pozzuoli uplift. Bull. Volcanol., 47(2): 187–200.

Bianchi, R., Coradini, A., Federico, C., Giberti, J.P., Sartoris, G. and Scandone, R. 1984. Modelling of surface ground deformation in the Phlegraean Fields volcanic area, Italy. Bull. Volcanol., 47(2): 321–330.

Bianchi, R., Coradini, A., Federico, C., Giberti, G., Luciano, P., Pozzi, J. P., Sartoris, G. and Scandone, R., 1987. Modelling of surface ground deformation in volcanic areas: the 1970–1972 and 1982–1984 crises of Campi Flegrei, Italy. J. Geophys. Res., 92: 14139–14150.

Buccheri, G. and Di Stefano, E., 1984. Contributi allo studio del Golfo di Pozzuoli. Pteropodi e Nannoplancton calcareo contenuti in tre carote: Considerazioni ambientali e biostratigrafiche. Mem. Soc. Geol. It., 27: 181–193.

Deino A.L., Orsi G., Piochi M., de Vita S. 2004. The age of the Neapolitan Yellow Tuff caldera-forming eruption (Campi Flegrei caldera – Italy) assessed by 40Ar/39Ar dating method. J. Volcanol. Geotherm. Res, 133: 157–170.

Di Filippo, G., Lirer, L., Maraffi, S. and Capuano M. 1991. L'eruzione di Astroni nell'attività recente dei Campi Flegrei. Boll. Soc. Geol. It., 110: 309–331.

Di Girolamo, P., Ghiara, M.R., Lirer, L., Munno, R., Rolandi, G. and Stanzione, D. 1984. Vulcanologia e petrologia dei Campi Flegrei. Boll. Soc. Geol. It., 103: 349–413.

Di Vito, M. A., Lirer, L., Mastrolorenzo, G. and Rolandi, G. 1987. The Monte Nuovo eruption (Campi Flegrei, Italy). Bull. Volcanol., 49: 608–615.

Marzocchi, W., Vilardo, G., Hill, D. P., Ricciardi, G. P., and Ricco, C. 2001. Common features and peculiarities of the seismic activity at Phlegraean Fields, Long Valley, and Vesuvius. Bull. Seismol. Soc. America 91: 191–205.

Oliveri del Castillo, A., and Quagliariello, M. T. 1969. Sulla genesi del bradisismo flegreo. Atti Ass. Geofis. Ital., Napoli, October 1–4.

Val d'Orcia, a Renaissance Agricultural Landscape

Alessandro Ramazzotti and Francesco Mauro

Val d'Orcia is the valley formed by the river Orcia in the southern part of Tuscany, in the Province of Siena, on the border with Latium. The river springs from a gorge on Mount Cetona, winds its way approximately northwestward crossing the municipalities of Cetona, Sarteano, Pienza, San Quirico d'Orcia, Castiglione d'Orcia and Montalcino, and is then tributary to the river Ombrone (in the Province of Grosseto) flowing toward the Tyrrhenian Sea. Thus, the valley runs between the hills south of Siena and the ancient extinct volcanic masses of Monte Amiata (1,738 m) and Radicofani. The geological form of the area was determined 5 million years before the present when the sea receded, leaving behind the sand and clay deposits that formed the valley surface. Later, the two volcanoes (the northernmost of a long line of extinct, dormant, or still active volcanoes on the Tyrrhenian side of Italy) covered the soil with erupted lava that, when it cooled, became a rock known as trachyte.

A pictorial map of the Val d'Orcia

A. Ramazzotti (✉)
Università Telematica "Guglielmo Marconi", Via Plinio 44, 00193 Rome, Italy
e-mail: alessandro.ramazzotti@macchind.it

N. Evelpidou et al. (eds.), *Natural Heritage from East to West*,
DOI 10.1007/978-3-642-01577-9_25, © Springer-Verlag Berlin Heidelberg 2010

Val D'Orcia, a World Heritage Site

The greater part of the valley has been recognized in 2004 by UNESCO as a World Heritage site for of its extraordinary environmental, landscape, and cultural features. In fact, the agricultural landscape of the valley has been gradually planned and modified for centuries, especially in the fourteenth and fifteenth century as a part of the territory of the republic (city-state) of Siena; this was an attempt to apply an idealized model of good governance and to create an aesthetically pleasing visual framework. The distinctive aspects of the natural landscape – flat chalk plains or

The Rocca (Castle) of Castiglione d'Orcia
from: http://www.flickr.com/photos/senpai27/2430928990/

The Rocca (Castle) of Tentennano in Rocca d'Orcia
(from a private collection)

gently rolling hills, occasionally broken by gullies, out of which rise higher hills topped by fortified settlements – inspired many artists, especially painters from the Siena School. Still, the dominant element of the landscape is represented by erosion processes that often reveal the argillaceous substrate, of a lighter color. In fact, erosion has played a major role in the formation of the landscape with the clay soil laid bare and forming craggy badlands known as "calanchi" and clay knolls otherwise called "biancane" or "mammelloni", that can be seen in the localities of Casa a Tuoma (Pienza), Ripalta (San Quirico), Lucciolabella, Beccatello, and Torre Tarugi (Pienza), Contignano, Pietre Bianche and the Poggio Leano (Radicofani).

Images of the Val d'Orcia, and particularly depictions of landscapes where people are shown as living in harmony with nature, have come to be seen as icons of the Renaissance and have profoundly influenced the development of thought about landscape. The valley has a variety of aspects: an agrarian and pastoral landscape reflecting innovative land management systems; isolated farmhouses; towns and villages; and the Via Francigena (that is, the Frankish Way, the north–south route followed by European medieval pilgrims and imperial expeditions to reach Rome using the ancient Roman route way of the Via Cassia) marked by abbeys, shrines, inns, bridges, towers, castles, etc.

The history of the valley is closely linked to that of the city of Siena. This was initially Etruscan territory, colonized by the Romans in the first century BC, part of a powerful Lombard and then Frankish marquisate during the high Middle Ages, in practice governed by the bishop, then by the Ghibelline (pro-imperial) commune and city-state of Siena from the twelth century until the conquest by Florence in the sixteenth century.

Landscape and Towns

The following municipalities, all situated in the Province of Siena, belong to the Val d'Orcia proper:

- Castiglione d'Orcia (142 km^2, 2,530 inhabitants);
- Montalcino (243 km^2, 5,130 inhabitants);
- Pienza (122 km^2, 2,230 inhabitants);
- Radicofani (118 km^2, 1,220 inhabitants);
- San Quirico d'Orcia (42 km^2, 2,525 inhabitants).

The land of these five municipalities form the Artistic, Natural and Cultural Park of the Val d'Orcia, a protected area whose aim is to preserve the natural and artistic heritage of the valley and, at the same time, improve the local economy and the way of life of the people, without turning the area into some sort of "museum". The Park is a "natural protected area of local interest" and is managed by a company, the Val d'Orcia s.r.l., whose members are: the five municipalities, the Province of Siena, the Comunità Montana ("mountain community consortium") Amiata Senese, individual businessmen, associations, consortiums, and other local authorities.

All these municipalities are interesting for art history, due to the medieval (eleventh to fourteenth century) and Renaissance age (fifteenth and sixteenth century) architecture. The most important sites are in:

- San Quirico d'Orcia (the Collegiate Church, Palazzo Chigi, and the Horti Leonini);
- Pienza (the Cathedral, the Church of San Francesco, the Pieve of Corsignano, the Ammannati, Piccolomini and Bishophric Palaces);
- Radicofani (the Rocca and the Medicean walls);
- Castiglione d'Orcia (the Rocca of the Aldobrandeschi and the Rocca d'Orcia);
- Montalcino (the Rocca, the Communal Palace, the Church of San Augostino).

Small villages scattered in the Val d'Orcia are also typical:

- Monticchiello, Corsignano, Castelluccio, and Spedaletto, in the municipality of Pienza;
- Rocca d'Orcia, Campiglia d'Orcia, Ripa d'Orcia, and Vivo d'Orcia, in the municipality of Castiglione d'Orcia;
- Vignoni, in the municipality of San Quirico d'Orcia;
- Castelnuovo dell'Abate, in the municipality of Montalcino.

Of great interest are the examples of religious architecture such as: the Camaldolese monastery of San Piero in Campo; the Olivetano monastery of Sant'Anna in Camprena; and – perhaps the most beautiful religious architecture site of all the Val d'Orcia – the Abbey of Sant'Antimo near Montalcino.

Some of these towns or hamlets are very picturesque and historically remarkable. Pienza has been purposely rebuilt in the fifteenth century as a planned "ideal town" under the patronage of Pope Pius II, Enea Silvio Piccolomini, born in the area in

The central square of Pienza
from: http://www.toscanaviva.com/Pienza

A farmhouse with cypresses near Pienza (from a private collection)

1405 and elected Pope in 1458. During his papacy, he changed the ancient Castle of Corsignano (first mentioned in 828) into a papal residence in the Renaissance style, constructed under the supervision of Bernardo Gambarelli called "il Rossellino", a student of Leon Battista Alberti, and renamed it Pienza (from "Pius"). Intended as a retreat from Rome, Pienza represented the application of humanist urban planning concepts, and initiated an attention to planning that was adopted in other Italian urban settings and thereafter spread to other European centers.

Radicofani is positioned on the pass of the Via Cassia leading toward the Latium and Rome. In fact, its main landmark is the Rocca (Castle), of Carolingian origin and documented from 978. It was the castle of Ghino di Tacco, a notorious robber baron (and popular hero, quoted in Dante's Purgatory and Boccaccio's Decameron) of the thirteenth century, operating at the border between Siena territory and the Papal States. Occupying the highest point of a hill, at 896 m, the Rocca was restored after the conquest by the Medici's Grand Duchy of Tuscany (1560–1567). It has two lines of walls: the external one has pentagonal shape, while the inner one is triangular, with three ruined towers at each corner and a "cassero" (donjon).

Montalcino is also famous as its red wine, the Brunello di Montalcino, is counted among the most prestigious Italian wines. Northwest of Bagno Vignoni, a magnificent rocky gorge, covered with woodlands and Mediterranean "macchia" (maquis), opens out onto the vineyards of Montalcino and then continues to the sea. On the slopes of Monte Amiata, there are forests of beech and chestnut trees; and of particular interest and rare beauty is the holm oak woods of Scarceta. The Abetina del Vivo, with ancient silver fir trees is situated near the old village of Vivo d'Orcia, famous for its springs which provide water for much of the area. Other trees found throughout the area are different species of oak with woodlands becoming thicker towards the swampy Maremma. However, the tree that has become a symbol of the Val d'Orcia is the cypress. The valley is also home to a large variety of small and medium size wild mammals and birds. Sheep farming is carried out in the Radicofani area.

A grove of cypresses in the Val d'Orcia (from a private collection)

Ancient SPAS

Due to its volcanic nature, the land is rich in hot springs, very often used for centuries as spas. The most remarkable is perhaps Bagno Vignoni on a hill just south of San Quirico d'Orcia; the name of this ancient village derives from Vignoni, already a famous castle in the eleventh century, whose remains dominate the hill over the village. These thermal waters were frequented already by the Romans, and the Renaissance spa is still intact next to the modern one. The spa was often visited by St. Catherine of Siena, pope Pius II, and Lorenzo de' Medici the Magnificent. A porch-type bridge passes over the waters flowing from the bath towards the thermal establishments and subsequently going on to feed a series of mills situated along the steep banks of the river.

The original spa of Bagno Vignoni
from: http://www.flickr.com/photos/lonelywolphoto/416417284/

The Commune of Siena always kept thermal treatments, carried out in its territory, under strict control and indeed two articles of the City Constitution are dedicated specifically to the Vignoni spa, prescribing the separation in the baths of men from women, with the cost of the operation to be borne half by the local inhabitants, together with the hotel keepers, and the other half by the inhabitants of the Val d'Orcia castles.

Other famous and pleasant spas are: Bagni San Filippo, on a steep hill side nearer Monte Amiata; San Casciano dei Bagni, adjacent to the valley, on the other side of Radicofani, with an intact historic center; and generally speaking, many other spas dotting the Province of Siena.

Some traditional "peasant" food can still be enjoyed: "pecorino" cheese (from sheep fed in pasture with aromatic herbs), aged by mixing it with walnuts or putting it under ashes, pig or wild boar salami (the famous "finocchiona" and the "sanguinaccio") and ham, some special pasta ("pici"), ox spleen croutons, pheasants "cacciatore", chestnut flour to make "castagnaccio", olive oil and red wine.

Conclusion

Nowadays, Val d'Orcia is a small historic region away from main communication routes and still unspoiled. The character of its agricultural economy (wheat, olive,

A farmhouse and a chapel near San Quirico d'Orcia
(from a private collection)

A view of the Val'Orcia
from: www.italytraveller.com/en/r/tuscany/s/val-d-orcia

grape and, traditionally, saffron), the persistence of the inhabitants in carrying out traditional work (carpentry, ironworks, pottery, terracotta, quarries), and the natural materials found in the area (timber, stones, clay), have allowed the preservation of a relatively balanced relationship between man and the environment.

As far as we know, Val d'Orcia is a very rare, relatively intact, example of Renaissance (and Medieval) landscape including all its human aspects. This is not an isolated phenomenon or an artificial reservation, but grows out of a long history, along the Via Francigena, starting in Etruscan-Roman times and translating appropriately into the modern era. Under this cultural layer there is a natural one: a scenery beginning with a symmetric harmony of soft rolling hills, interspersed with creeks, ravines, outcrops, river banks, changing into the grand greenery of Mount Amiata. The solitude, the empty spaces, the light, the views, induce a sense of pleasure, delight, and even loneliness difficult to define.

It is a unique case where nature and culture, history and geography, landscape and art, come together to illustrate the good side of human activity.

Websites

en.wikipedia.org/wiki/Val_d'Orcia
whc.unesco.org/en/list/1026
www.nautilus-mp.com/tuscany/presentazione/valdorcia/indexing.html
www.sienaonline.com/val_d_orcia_tuscany.html

Vallo di Diano: A Highland in South Italy

Francesco Mauro

A little known part of Italy: a picturesque high valley where Campania meets Basilicata and Calabria, at a crossing point of communication, peoples, history, culture and arts.

Geography and History: The Valley as a Gateway

The Vallo di Diano (Dianum Valley) is a high valley (450 m average elevation) in the southeast part of the Province of Salerno in the Region of Campania, Italy.

The geography of the Vallo di Diano is well illustrated in the map below, modified from an official map of the Cilento and Vallo di Diano National Park, where the middle of the valley, on the right of the map, corresponds to the indicated expressway.

The valley is bordered by two mountain chains: the Monti Alburni (maximum elevation Monte Panormo 1,742 m) on one side and the Monti della Maddalena on the other, on the border with Basilicata. Going from west to east, we find:

- the lower Tyrrhenian coastline south of the Gulf of Salerno, very rocky, with rugged cliffs, difficult to follow;
- the large mountainous area of Cilento;
- the Vallo di Diano, with the Tanagro river running through it northwest to the Sele river and to the Tyrrhenian Sea;
- and on the other side of the Monti della Maddalena, the Val d'Agri (in the Basilicata Region), with the Agri river running southwest to the Ionian Sea.

This latter orientation has been fundamental in determining the historic destiny of the valley. In a section of the Italian peninsula where the coast was unfriendly and unusable, it offered a practicable means of inland communication between the north (Rome, Capua, Naples and Salerno, that is, the main part of Italy and Europe)

F. Mauro (✉)
Università Telematica "Guglielmo Marconi", Via Plinio 44, 00193 Rome, Italy
e-mail: mauro_sustainability@yahoo.com

N. Evelpidou et al. (eds.), *Natural Heritage from East to West*,
DOI 10.1007/978-3-642-01577-9_26, © Springer-Verlag Berlin Heidelberg 2010

and the south (Calabria, Sicily and, ultimately, Africa). Furthermore, the proximity of Val d'Agri and the lack of difficulty of the mountain passes, allowed an easy route crossing the Apennines to Basilicata, and from there to Apulia, the Ionian and southern Adriatic coasts of Italy, Greece and the East.

This natural passage and crossroad has been well traveled by humans, as a route for prehistoric and protohistoric man, of Italic (Oenotrian and Lucanian) and Illyric tribes, and also as a path for commerce between the Etruscan-Roman and the Greek

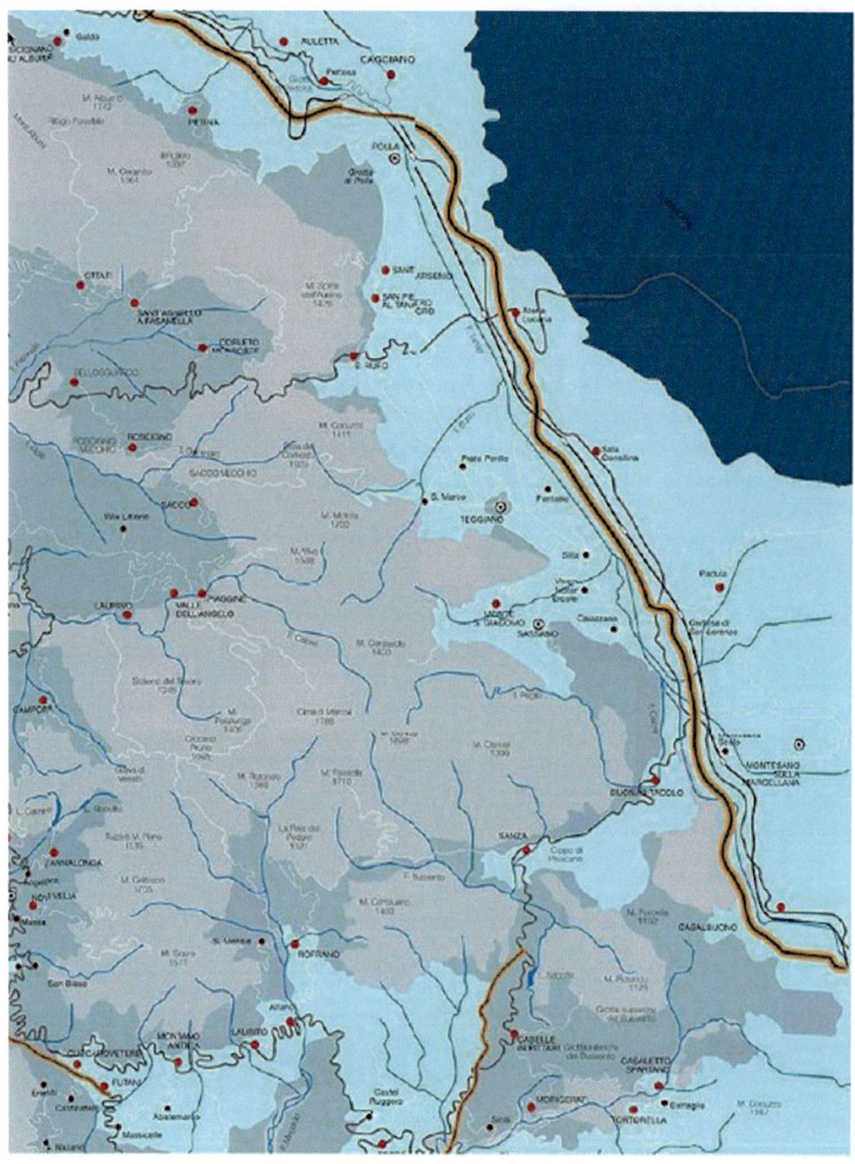

A map of the Vallo di Diano (see text)

sphere of influence, between the Tyrrhenian and the Ionian cities of Greater Greece, that were to be conquered by Rome. In 128 BC the Romans built in the Vallo di Diano the military road Via Annia or Popilia. The road was used by the Visigoths of Alaric in the invasion of 410 AD, and thereafter by Longobards and Byzantines (forming a contact point between West and East), Normans, Swabians, Anjevins, Aragonese, Spaniards, Frenchmen, Neapolitans, right up to the unification of Italy by Garibaldi in 1860. Nowadays the old Roman road is still there, kept up by Bourbon kings, Napoleonic regimens and by modern united Italy. It is now known as the national route 19 "of Calabria"; and next to it, along the Tanagro, the A3 expressway to Reggio Calabria, confirming the easiness of the passage. On the coast, served by a railroad with many tunnels, there are only minor and tortuous roads.

The Vallo di Diano is an alluvial valley, an ancient Pleistocene lake. The flat part of the valley has been traditionally used for farming and has recently experienced some urbanization, often unfortunate and sometime illegal. Nevertheless, the mountain landscape is relatively intact and is part of the Cilento and Vallo di Diano National Park, the second largest of Italy.

The area has a mild climate but is unfortunately exposed to seismic and hydrogeological risk, for instance the Irpinia earthquake of 1980, landslides and floods.

The Vallo di Diano is administered as 15 communes (municipalities), united in a "mountain community" (comunità montana), with a total of some 65,000 inhabitants (this figure becomes 100,000 on working days when commuters come from surrounding areas). Sala Consilina has almost 15,000 inhabitants; Polla, Montesano sulla Marcellana, Teggiano, and Sassano over 5,000; Atena Lucana (with an industrial area), Padula, Casalbuono, Sanza, Buonabitacolo, Monte San Giacomo, San Rufo, San Pietro al Tanagro, Sant'Arsenio less than 5,000. The smallest commune is Pertosa.

A view of the Vallo di Diano looking west from the hill where Padula is located. In the foreground: the Chartreuse. Note the flat alluvial soil in the middle of the valley

Cultural and Natural Heritage Sites

The area is rich of cultural heritage sites and structures. First of all, there is the Chartreuse (Certosa) of San Lorenzo in Padula, a UNESCO world heritage site together with the ruins of Greek-Roman Paestum and Velia (outside the Vallo) and the National Park. Certosa is a large famous Carthusian monastery, the second largest in Italy after the one in Parma. It was founded by the Norman feudal lord, prince Tommaso Sanseverino, in 1306, on the site of an existing coenobitic (hermit) cell. It is dedicated to St. Lawrence, and its architectural structure recalls the grate upon which the martyr saint was burnt alive. The main parts of the building are now in Baroque style. It is a large complex: 51,500 m^2 (12.7 acres) in all with 320 rooms and halls; it has the biggest cloister of the world: 12,000 m^2 (2.97 acres), surrounded by 84 columns. A famous spiral staircase of white marble inside an annex leads to the large library. It is now open to visitors.

The grand cloister of the Chartreuse of Padula

Other natural and cultural heritage sites are:

- the Angel Caves of Pertosa, a spectacular karst complex, formed 35 millions years ago, of scientific, historical, and tourist interest;
- the important and delightful Archeological Museum of Western Lucania, housed in a section of the Chartreuse, and rich in pre-Roman items;
- the historical centers of the towns of Teggiano (the museum-town once known as Dianum, the ancient capital of the Vallo), Auletta, Polla, Sala Consilina, Padula (with the paleochristian Baptistry of San Giovanni in Fonte located between the last two towns), etc.;

- the so-called "valley of the orchids" near Sassano;
- the Greek-Roman archaeological area between Sassano and Monte San Giacomo;
- some ten small museums, a diffuse network in the different towns, dealing with archaeology, speleology, prehistory, ecology, botany, folk tradition, and rural civilization.

The grand cloister of the Chartreuse of Padula with the town of Padula in the background

Downtown Teggiano: the palace-castle of the princely Sanseverino family

Traditional Production

The Vallo is famous for the production of traditional foodstuff such as the Pertosa's white artichoke, peppers, eggplants, some varieties of figs, olive oil, some typical sausages and salami, cheeses (including mozzarella, scamorza and caciocavallo), durum wheat flour pasta, etc. Aglianico red wine is popular.

Some traditional craftsmanship is still present: stone carving, leather, wood carving, embroidery and laces (especially for bridal dressing), artistic wrought iron working, wicker objects. In the area there is an ongoing crisis of non-specialized agriculture and manufacturing industry, but recently an effort is been dedicated to increasing the value of sustainable tourism, using the Chartreuse a main point of attraction: craftsmanship, traditional products, and advanced cultural and tourist services may well complement this trend.

Websites

http://whc.unesco.org/pg.cfm?cid=31&id_site=842
http://www.parks.it/parco.nazionale.cilento/Eindex.html
http://www.ultimateitaly.com/unesco/certosa-di-padula.html

Valnerina: The High Valley of Hermits, Friars and Saints

Francesco Mauro and Ilaria Reggiani

Valnerina: A land of nature and culture, of abbeys and convents, of solitude and banquets, of earthquakes and peace, in the center of Italy.

Valnerina is the narrow valley formed by the river Nera – one of the main tributaries of the Tiber, the river of Rome – located in the southeastern part of the

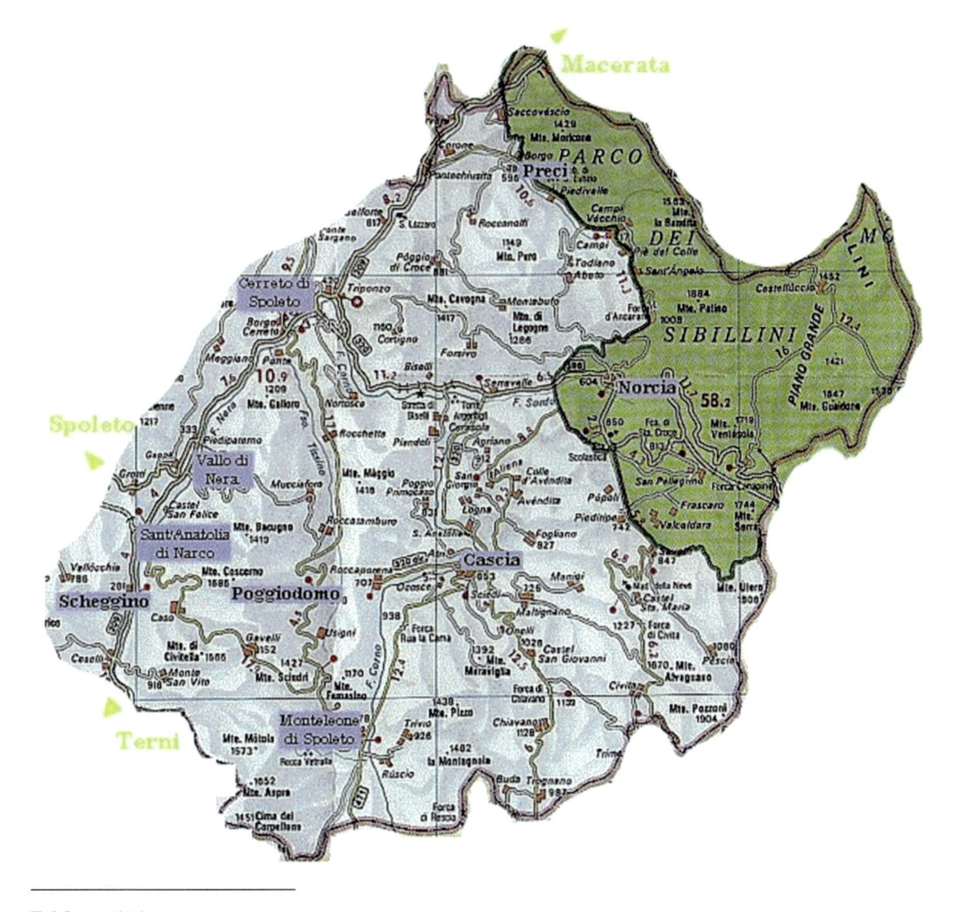

F. Mauro (✉)
Università Telematica "Guglielmo Marconi", Via Plinio 44, 00193 Rome, Italy
e-mail: mauro_sustainability@yahoo.com

N. Evelpidou et al. (eds.), *Natural Heritage from East to West*,
DOI 10.1007/978-3-642-01577-9_27, © Springer-Verlag Berlin Heidelberg 2010

Region of Umbria, approximately in the center of the Italian Peninsula. The area – the main valley and its secondary valleys – is in part under the administration of the Province of Perugia (the "communes" or municipalities of Cascia, Cerreto di Spoleto, Monteleone di Spoleto, Norcia, Poggiodomo, Preci, Sant'Anatolia di Narco, Scheggino, Sellano, Spoleto and Vallo di Nera) and in part under that of the Province of Terni (the municipalities of Arrone, Montefranco, Ferentillo, Polino e Terni).

Brief Description

The Valnerina is a typical narrow and steep valley of the Apennine Mountains, running between tablelands and naturally divided into two sections: the lower valley, corresponding to the Province of Terni, and the upper valley, corresponding to the Province of Perugia. The latter, the Alta Valnerina, is organized as Comunità Montana della Valnerina (mountain community inter-municipal consortium formed by 10 communes), part of which (Norcia and Preci) lies in a national park, Parco Nazionale dei Monti Sibillini. The area of the upper Valnerina is 918 km^2, and the, decreasing, population is 13,766, with some 5,000 in Norcia.

The valley of the river Nera, flowing into the Tiber further south at Terni, is widely considered the most scenic area of Umbria, "il cuor verde d'Italia" (the green heart of Italy). While the upper Nera flows more or less by itself in the mountains, the lower basin is a fairly large floodplain, which in ancient times contained a pair of shallow interlocking lakes, the Lacus Clitorius and the Lacus Umber; these were drained by the Romans after interventions lasting several hundred years, and re-drained in the thirteenth and eighteenth centuries.

In the lower part of the Valnerina, an outstanding feature is the Cascata delle Marmore, the highest water fall in Europe (165 m). This is also of artificial origin, being built in 271 BC by order of the Roman consul Curius Dentatus to control flooding in the marshy Sabine valley by directing the waters of the river Velino into the river Nera. In modern times, the river course, at the point of the falls, has been partially diverted to supply hydroelectric plants serving several industries (now in large part closed down).

The valley, although it is rural, scarcely populated, relatively isolated and protected, is very close to landmarks of the Italian civilization. The area has been papal territory from the fifteenth century to the unification of Italy in 1862. Prior to that, Spoleto (the city outside but adjacent to the valley) had been the capital of a high Middle Age powerful Lombard duchy and for a very short time in the ninth century it was even the seat of the German emperor and king of Italy. Nowadays, it is the main venue, together with Charleston (South Carolina), of the famous Festival of the Two Worlds of performing and visual arts. Assisi, the city of St. Francis and his Cathedral with the Giotto's frescoes, is nearby. Cities of Etruscan origin (Perugia, Todi, Orvieto, etc.) are also close; Terni, once upon a time seat of bishop Valentine (patron saint of lovers), has been an important modern industrial center of iron and steel production.

Valnerina and the Religious Influence

The development of the area has been deeply affected, if not triggered, by the presence of religious communities attracted by the isolation and the peace of the valley. First, in the fourth to fifth centuries, Syriac anchorites escaping persecutions and invasions settled in isolated gorges and on peaks. Then, from the sixth century, with the growth of Western Christian monastic organization, there were the followers of St. Benedetto da Norcia (Benedict of Nursia, 480–547, who was born in the valley) and of his sister St. Scholastica. This group, the Benedictine Order, was extremely important for the preservation of culture in difficult times and the development of the idea of Europe.

The monks of the numerous abbeys located in the valley were essential in defining the agricultural landscape. As in later times and at other places, such as Montecassino, they modified the territory by wood clearing, stone removal, introduction of wheat farming, pastoral practice, planting of olives and orchards, drainage and reclamation of marshes (the "Marcite" on Norcia's karst plateau). The economy of the valley is still based not only on tourism and commerce, but also on quality agriculture (including niche products such as Castelluccio lentils) and pig meat processing. In the modern Italian slang typical of Rome, a person employed in traditional sausage and prosciutto ham preparation, is sometimes called a "norcino" (from the town of Norcia).

The next wave of development by a religious community took place in the twelth to thirteenth century. The valley, so close to the city of St. Francis of Assisi, became part of the Franciscan heartland. Franciscan churches and convents were gradually

An isolated church with a small convent

added to Benedictine abbeys to create a network of civilian and sacred edifices, connecting roads, and landscape improvements. These are still distinctive elements of the valley.

Another relevant episode, starting in the fifteenth century, is linked to the figure of St. Rita of Cascia who was born in the valley, in Roccaporena di Cascia. In the second part of her life she was a member of a very ancient order, the Contemplative Augustinian Nuns, at their convent in Cascia. Due to her difficult but exemplary life, first as a mother of a family and then as a pious nun, she became an extremely popular Catholic saint and her sanctuary (built after her quite recent canonization in 1900) in Valnerina is one of the most frequented pilgrimage sites in Italy.

Features of the Valnerina

Even now, the Valnerina maintains its original authentic features. The territorial structure has been preserved, even in its minor aspects. From these features (water-control systems, monasteries and isolated churches, road network, farms, urban settlements, villas and castles, and modern industrial archeology sites) developments of different periods, from ancient to modern times, can be easily recognized.

A strong earthquake struck the Apennines in Umbria and Marche in September–October 1997. This was a sequence of seismic events over approximately one month that caused serious damage to several monuments (including Assisi Higher Basilica) and minor edifices. Following this event, all the buildings involved have been restored and repaired or consolidated under the supervision of the competent authorities. Unfortunately, the level of seismicity is high in Umbria, along the very active Umbria-Marche fault alignment; and the Valnerina is particularly susceptible. The towns in tributary valleys, and especially Cascia and Norcia, have suffered dramatic earthquakes: a catastrophic one in the first century BC (destroying Carsulae = Cascia), then others in 1279, in the fall-winter of 1300, in November 1599, and in January 1703, with more than 700 victims. In the event of 1997 the earthquake-prone area of the Valnerina was on the southern margin of the activated zone, but it was more affected by the other seismic sequence of 1979 when higher amounts of energy was probably discharged. In the period 1703 to 1979, and again 1997, 8 earthquakes triggered landslides and hydrogeological changes in central Valnerina.

Another distinctive element of the Valnerina is the preservation of large portions of the original Roman water-control system, with the Cascata delle Marmore as the most significant example. The preservation of a dense network of civilian and sacred edifices built during the Middle Ages around the three main abbeys, further distinguishes the Valnerina. In particular, the original features of the two types of settlements established in this area, villas and castles, have been maintained here but can no longer be found elsewhere.

Ancient waterworks in the Valnerina

As far as tourism is concerned, the area is famous for architecture, cultural heritage, natural mountain trails, gastronomic routes (for sausages, boar meat, black truffles, etc.), rafting and canoeing, and devotional pilgrimages; but is not overcrowded – a fraction of the housing is unoccupied – and still relatively intact in terms of natural and cultural heritage. The objective of sustainable development and tourism is discussed and promoted.

The central square of Norcia

From 1926 to 1968 a local electric railroad, 51 km long, linked Spoleto (on a main rail line and the ancient Via Flaminia) to Norcia, serving a good section of the upper Valnerina by means of a daring route of tunnels, bridges, and viaducts. Unfortunately, the rail service has been discontinued, but the path and infrastructures are still there. This suggests the creation of a greenway, if not a tourist railroad.

Valnerina, together with the Cascata delle Marmore, is a candidate to be recognized as a World Heritage Site by UNESCO.

A viaduct of the Spoleto – Norcia railroad line (now closed down)

Websites

http://en.wikipedia.org/wiki/Umbria

http://penelope.uchicago.edu/Thayer/E/Biographical/Diary/edited/0900/summary.html

http://whc.unesco.org/en/tentativelists/2031

http://www.roughguides.com/website/travel/destination/content/default.aspx?utitleid=21&xid=idh396172192_0653

Volcanoes and Crater Lakes in Latium: Nature and History of Rome

Francesco Mauro

Four volcanic districts, marked by beautiful crater lakes, dot Latium north and south of Rome: they are an important natural feature affecting the history of Rome.

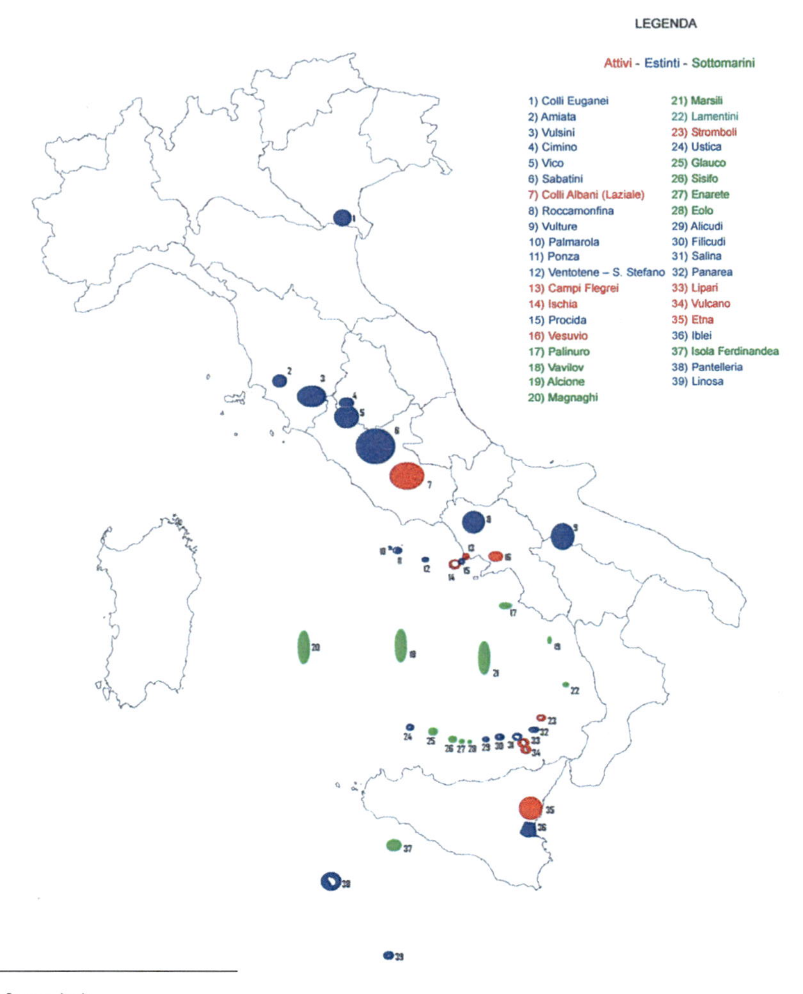

F. Mauro (✉)
Università Telematica "Guglielmo Marconi", Via Plinio 44, 00193 Rome, Italy
e-mail: mauro_sustainability@yahoo.com

N. Evelpidou et al. (eds.), *Natural Heritage from East to West*,
DOI 10.1007/978-3-642-01577-9_28, © Springer-Verlag Berlin Heidelberg 2010

Italy is known also as a land of volcanoes. Some of them are well known, described in popular books and motion pictures, even employed in scientific terminology: Vesuvius, in the area bordering the Gulf of Naples in the Tyrrhenian coast, protagonist of the famous eruption in 79 AD causing the destruction of Pompei, Herculaneum and Stabia; Stromboli, one of the Aeolian Islands in the lower Tyrrhenian Sea, a volcano exhibiting a persistent activity with continuous volcanic emissions; Vulcano, another of the Aeolian Islands, where the very name of the fire mountains originated; and Aetna, in Sicily, the largest subaerial emerged volcano in Europe. But, in fact, Italy boasts at least 40 volcanoes, including those active, dormant and extinct, on land, on volcanic islands and submarine. In particular, there is a string of volcanoes, parallel to the Tyrrhenian coastline, that, starting in Campania (where Vesuvius is located), enters into Latium, crossing it from southeast to northwest through four distinct component volcanic districts, and continues into Tuscany.

These volcanic districts of Latium are:

- the Colli Albani (Alban Hills), south of the Tiber valley where Rome is located;
- the Monti Sabatini (Bracciano), north of the Tiber valley;
- the Vico-Cimino system, between the Monti Sabatini and the Viterbo plain;
- the Monti Volsini, north of Viterbo, with the largest lake (Bolsena).

It has been noted that the position, the foundation, and the growth of Rome has been governed by geography and geology. In fact, the Tiber river runs between two of these volcanic mountain systems: thus, it offered a communication waterway from the sea to the interior of the Italian peninsula (a trade route for the Etruscan-Roman-produced salt and for the Greek artifacts destined to Italic tribes); at the same time a ford, and later a bridge, at the Tiberina Island allowed a route from the northern Etruscans to the southern Etruscans, the Greek colonies of Campania, the rest of south Italy and the East, to cross the river. The enormous quantity of material erupted from the two volcanoes, Albano and Sabatino, provided abundant tuff and lava, The former is a very strong, light and easily cut stone, of which Rome was made, and the latter was excellent for paving Roman roads. Another product of vulcanism was pozzolana, essential to make cement. Thus it was in this situation that Rome was founded, on the traditional date of April 21 of the year 753 BC.

The Alban Hills

The Alban Hills, also known as the Colli Albani or “i Castelli Romani”, from the Castles that dominate each local town, lie on the southeast of Rome. The volcanic system of the area, the Vulcano Laziale, forms a prominent feature. The structure is complex: the volcano, of the Somma-Vesuvius type, has two nested semi-collapsed calderas (the larger/older one is called “Artemisia-Tuscolana”) and numerous more or less eccentric post-caldera vents, most of which are explosion craters. The highest point is Monte Cavo (949 m), a scoria cone sitting eccentrically on the rim of the younger "Faete" caldera. The intra-caldera space is called "Campi di Annibale" (Hannibal's Fields): the legend, probably untrue, says that from here Hannibal, the Carthaginian general super-enemy of Rome, has been observing the city without daring to attack it.

There are now two major crater lakes, Albano and Nemi, which fill the most recent explosion craters of the volcano, and also some minor crater lakes, and lakes drained by the Romans (including the very large Aricia crater) as well. Lake Albano (also known as Lake Castelgandolfo), with its surface level at 293 m elevation, is

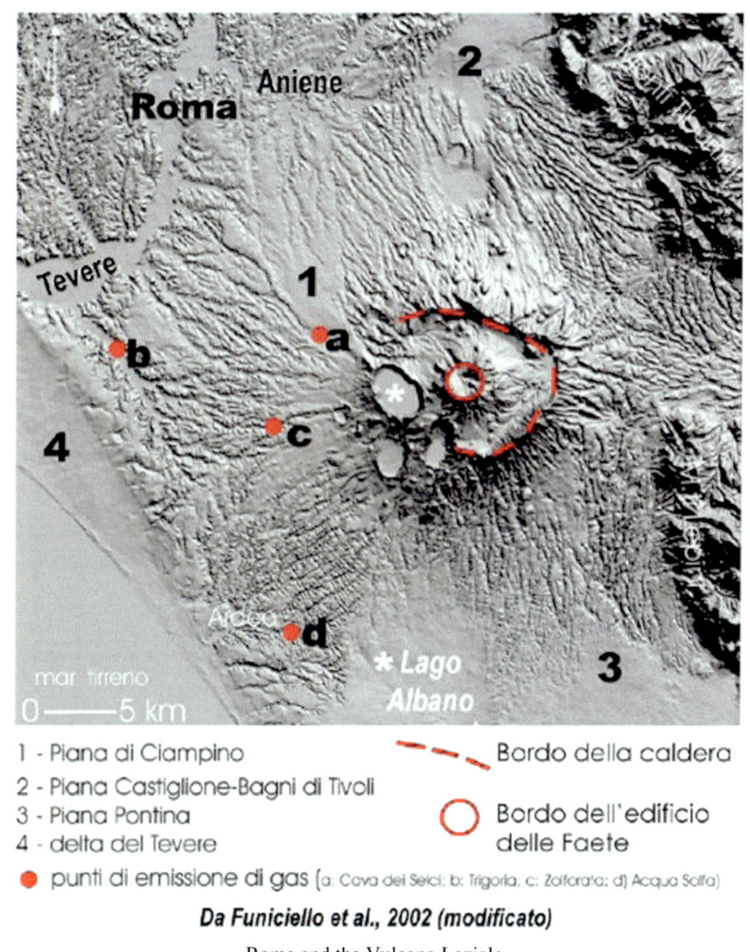

Rome and the Vulcano Laziale
From http://www.conoscoimparoprevengo.net/articolo.php?pid = 35&month = 4

1, 2: plains formed by volcanic material. 3: alluvial plain. 4: Tiber delta. a, b, c, d: current gas emissions. Continuous line: older caldera (Tuscolano-Artemisia). Dashed line: younger caldera (Faete). A number of craters (Albano, Nemi, Aricia, smaller ones) can be seen.

170 m deep and thus the deepest of all volcanic lakes in the Central Italian volcanic region. It is well known for its artificial outlet, which reaches a total length of 1,425 m as a tunnel under the crater rim. According to Livy, the outlet was built in 398–397 BC to control catastrophic changes of the water level of the lake, and it probably made use of a previously existing tunnel of the sixth century BC. Lake Nemi lies somewhat higher (316 m) but is very shallow. Lake Nemi was called Speculum Dianae, the "mirror of Diana", holy to the goddess, and host to mystery cults and to luxurious pleasure boats. Today, both lakes present are very scenic. They host the summer residence of the Pope at Castelgandolfo, right on the rim of the crater containing Lake Albano, very close to the site of Alba Longa, the mythical ancestral city of Rome.

The volcanic complex is relatively densely populated, with two circles of castle-towns nested on it, and the outskirts of Rome extending to its northern base. The area of the Castelli occupies this fertile volcanic area, which has, since ancient times, supported a flourishing agricultural economy and allowed the production of famous white wines (Frascati, Marino, Lanuvio, etc.). From the Roman era onwards the area has been frequented by the noblemen of Rome because of its fresher climate during summer in a malaria-free environment. The area is full of Roman villas, Medieval fortifications, Renaissance gardens and palaces. Aristocatic families that ruled in the Castelli include the Orsini, Colonna, Savelli, Chigi, Aldobrandini and Ruspoli.

Much of Rome itself stands on the products of the Pleistocene explosive volcanic activity of the Alban Hills. The 7 hills of Rome are made up of Albani eruptive material. The Appian Way runs on top of an extremely long solidified lava flow. In spite of frequent seismicity, the Alban Hills have, until recently, been considered an "extinct" volcano, but some (contested) historical documents indicate some kind of eruptive activity as recently as 114 BC. Livy writes (Book XXII): "et Praeneste

ardentes lapides caelo cecidisse" ("and at Praeneste there had been a shower of red-hot stones") in about the year 216 BC in a locality on the eastern side of the volcano. Since the mid-1990s, increasing evidence for ongoing volcanic unrest has been detected, and the most recent research indicates that there was eruptive activity during the Holocene: thus, the volcano might be at the beginning of a new eruptive phase.

The presence of a potentially active volcano at only 25–30 km from the center of Rome should be taken seriously and should stimulate revised views of volcanic hazards in the area.

The Monti Sabatini

North of Rome, the Lake of Bracciano is the largest structure of the Sabatini complex. The lake fills a volcano-tectonic depression formed about 370,000 years ago. The Sabatini volcanic system (Vulcano Sabatino) is probably the most complex of the region. The various explosive centers of the complex developed on a vast plain formed of Plio-Pleistocene clayey and sandy sediments.

The first eruptive activity was mainly explosive and occurred in the eastern part of the complex, building a volcanic edifice named "Morlupo – Castelnuovo di Porto". While activity was ongoing there, eruptive activity also began towards the west, from what is probably one of the most important eruptive centers of the Sabatini complex, the "Sacrofano" edifice. Its activity spanned a long period, from 600,000 to 37,000 years ago, producing large volumes of ash flow tuffs. The pyroclastic flows from the Sacrofano center reach distances of 30–40 km from their source, covering a large portion of what is now the northern part of Roma. These tuffs yielded (and still yield) the primary material for building construction in the city.

It is presently believed that the culminating phase of Sabatini volcanism was related to high stress connected with extensional tectonics that controlled the evolution of the Tyrrhenian margin of the Apennine peninsula during that period. Due to tectonic activity and simultaneous emptying of magma reservoirs, there ensued the collapse of the depression now occupied by Lake Bracciano and the subsidence of the structural high of Baccano-Cesano. Following the end of Sacrofano activity, minor eruptive activity occurred from vents in the eastern part of the complex. The tuff rings of Monte Razzano and Monte Sant′Angelo as well as the complex center of "Baccano" were formed during this period. The last Baccano eruptions occurred 40,000 years ago, but, more recently, small-scale eruptive episodes occurred from the Martignano (three overlapping craters), Stracciacappe and Cese eruptive centers. Martignano is now a lake, some crater lakes have been drained (Baccano and Stracciacappe), one crater (Manziana's Caldara) still emits gases and sulfurous water, and another lake (Monterosi) is not volcanic.

Lake Bracciano, 32 km northwest of Rome, with a surface of 57 km^2, is the second largest lake in the region and one of the major lakes of Italy. It has a circular perimeter of approximately 32 km; its surface is 160 m above sea level and its deepest point is 165 m. The lake covers more than a single craters; the shape of some minor craters can till be recognized overlapping the major rim. The lake has one outlet, the Arrone River, running to the Tyrrhenian coastline north of the Tiber delta.

Three towns border Lake Bracciano: Bracciano (with a famous, well kept castle), Anguillara Sabazia and Trevignano Romano. This part of Latium, known as Lower Tuscia, is already in the heart of Etruscan land: the ruins of Veii, the first serious competitor of Rome, Tarquinia, and Ceri are nearby.

The area was dominated from the Middle Age by the Anguillara-Orsini family, challenged in the Renaissance by the Borgia papal family, and acquired by the Odescalchi in later times.

Map of Lower Tuscia: the Bracciano Lake (with the smaller Martignano Lake) and Vico Lake

The lake is an important tourist attraction situated near Rome. It serves as a drinking water reservoir for the city, the use of motorboats is strictly forbidden, and a circular sewer system with a centralized treatment plant has been built for all the bordering towns in order to avoid any degradation of water quality.

In the last few years, the lake and its surroundings have been brought under further protection by the creation of a regional park, the *Parco Regionale del complesso lacuale di Bracciano-Martignano*.

The Vico and the Cimino Systems

In contrast with the other volcanoes in Latium, Vico has a relatively simple structure, constituted by a central strato-volcano truncated by a caldera which in turn contains a post-caldera cone. The volcanic history of Vico spans from 800,000 until about 85–90,000 years ago. Four phases of activity can be described. During the first, magma was erupted from minor eruptive centers to form numerous air fall units and lava flows. Essentially effusive activity of a central volcano characterized the second period. The third phase was much more explosive with cataclysmic Plinian eruptions that were accompanied or followed by collapse 150,000 years ago of a 7.5 km diameter caldera which is now filled with the beautiful Vico Lake. The fourth and final phase of volcanism at Vico was strongly influenced by the presence of a lake filling the newly formed caldera, and thus was violently hydro-magmatic. Only in its very final stage the character of the eruptions became magmatic again, building 95,000 years ago the eccentric cone of the Monte Venere (851 m).

The Vico Lake is the most elevated among major Italian lakes, with an altitude of 510 m. Administratively, it is part of the municipalities of Caprarola and Ronciglione. According to the legend, the lake was created by Hercules, who had defied the local inhabitants to pick up his club; when he did it by himself, a stream sprung on the spot and formed the lake. Before the construction of a tunnel by the Etruscans to control the water level, the lake had probably a surface twice than today, the Monte Venere being an island within it. The lake has now one outlet, the Rio Vicano. It is famous for its extensive beech forest, one of the most southerly in Europe. The elevation, together with the covering effect of the surrounding crater rim, apparently supplies cold enough conditions for the survival of the beech trees. In order to preserve this forest, a large part of the northern side of the crater is a natural protected area.

Very close to the Vico volcano there is the Monte Cimino (1,050 m), the oldest of the volcanoes of the region (active from 1,350,000 to 800,000 years ago). The growth of many domes was accompanied by violent explosive activity and probably by collapse and avalanching from the domes, both of which generated glowing avalanches. More than 50 lava domes are still recognizable in the Cimino area, and many more are supposed to either lie buried below younger domes and their pyroclastic flow aprons or to have been annihilated by final explosions. The domes that are still present give the Cimino area its characteristic hilly morphology.

The Vulsini Volcanic Complex (Vulcano Volsino)

Is the northernmost of the volcanic districts in Latium. Its activity has been principally explosive. The main structural element is the vast basin of the Bolsena Lake, a volcano-tectonic depression that has formed during successive phases of subsidence. The activity of the complex has occurred from four main eruptive centers on the margins of the main depression, probably along the main tectonic lines of weakness. Activity probably initiated about 800,000 years ago; about 600,000 years ago, the activity concentrated at a center denominated "Paleovulsini" in the area now occupied by the lake depression. Another important eruptive centre then grew immediately and is called "Bolsena-Orvieto". Contemporaneously with the Bolsena-Orvieto center, another center named "Montefiascone" was active at the site now occupied by the town of Montefiascone (300,000–150,000 years ago). Still another major eruptive centre was active during that same period, on the west side of the Bolsena depression, named "Latera".

Bolsena Lake is the crater lake of the Vulsini volcanic complex. Roman historic records indicate that activity of the Vulsini volcano occurred as recently as 104 BC, since when it has been considered dormant. The two islands in the southern part of the lake have been formed by underwater eruptions following the initial collapse of the caldera. The lake has an oval shape with a total surface is 113.5 km^2; the elevation of its surface is 305 m; it is 151 m deep at its lowest point and 81 m deep on average.

The lake lies within the northern part of the Province of Viterbo that is called Tuscia. It is bordered on the western and northern side by the Roman consular road Via Cassia (the medieval via Francigena used by the pilgrims going to Rome). The Bolsena Lake has numerous lodging establishments, particularly for nature-oriented tourism, camping and agro-tourism. The Romans called it Lacus Volsinii, adapting the Etruscan name, Velzna, of the last Etruscan city (the sacred capital of the 12-cities Etruscan confederation) to hold out against Rome, destroyed in 264 BC.

With an area of 17 ha, Bisentina is the largest island, characterized by oaks, Italian gardens, enchanting panoramas and numerous monuments, including the *Malta dei Papi*, the prison for life for ecclesiastics found guilty of heresy. Located opposite the town of Marta, the smaller island of Martana is reputed to have been the prison of the Gothic queen Amalasuentha who met a horrible death there after a dynastic conspiracy.

The Marta outlet leaves Bolsena Lake and, after passing through Marta, Tuscania and Tarquinia, reaches the Tyrrhenian Sea. A number of communes (municipalities) borders the lake: the main ones are Bolsena and Montefiascone.

Conclusion

The Latium volcanoes are not only a relevant aspect of the landscape of Rome and its territory, but they played an essential role in determining the origin and evolution of the city and the Roman state. The Roman landscape is unique: cones and domes covered by woods, fertile volcanic soils but also small canyons and badlands created by erosion, tuffs, crater lakes, Mediterranean bush and, together with these natural features, Etruscan and Roman sites, necropolises and ruins, aqueducts and canals, castles and tower, Renaissance and Baroque villas and palaces, churches and monasteries, villages and towns, and at the center of this land, between the sea and the mountains, between two volcanic complexes, the *Urbs* (the City). This is a unique blending of nature and history, where natural risks may at the same time represent past and current assets for human development and culture.

Websites

http://consar0.startlogic.com/lakes10.html
http://vulcan.fis.uniroma3.it/misc/miths.html
http://www.volcano.si.edu/world/volcano.cfm?vnum=0101-003
http://www.volcano.si.edu/world/volcano.cfm?vnum=0101-004&volpage=synsub

References

De Rita D. 1993. Il vulcanismo. In: Società Geologica Italiana (ed.) Guide Geologiche Regionali: Lazio: 50–64.
De Rita D, Funiciello R, Rossi U and Sposato A. 1983. Structure and evolution of the Sacrofano-Baccano Caldera, Sabatini Volcanic Complex, Rome. Journal of Volcanology and Geothermal Research 17: 219–236.
Kilburn C, McGuire B. 2001. Italian Volcanoes (Classic Geology in Europe 1). Terra Publishing.

The Natural Heritage of the Island of Gozo, Malta

Dirk De Ketelaere, Anna Spiteri, and Josianne Vella

The name "Gozo", meaning "joy" in Castilian, was given to this small island by the Aragonese who took over the Maltese Islands in 1282. Over the years the island has inspired many names. In 700 BC the Phoenicians called it "Gwl" or "Gaulos", meaning round ship, a name that was retained by the Greeks and translated by the Romans as "Gaudos" or "Gaulum". The Moors, who ruled the Maltese islands around a thousand years ago, and who strongly influenced its Semitic language, referred to it as "Gaudoich" preceding the current name *Għawdex* (pronounced Aw-desh), which is used today by the local inhabitants. In a legendary context it is often called the "Island of Calypso" referring to the Greek mythological location of Ogygia, home of the beautiful nymph Calypso. In Homer's epic poem, The Odyssey, Calypso keeps the Greek hero Odysseus as a prisoner of love for 7 long years.

Introduction

The Maltese Archipelago, located at the centre of the Mediterranean Sea, consists of the islands of Malta, Gozo and Comino as well as a few other uninhabited islets. Gozo is the second largest island with a coastline of 47 km, and a surface area of 66 km^2. The Islands have a typical Mediterranean climate, with mild, wet winters and long, dry summers. The average annual rainfall amounts to around 530 mm and follows a clearly marked seasonal rhythm.

D. De Ketelaere (✉)
Integrated Resources Management (IRM) Co. Ltd. Malta
e-mail: info@environmentalmalta.com
website: www.environmentalmalta.com

N. Evelpidou et al. (eds.), *Natural Heritage from East to West*,
DOI 10.1007/978-3-642-01577-9_29, © Springer-Verlag Berlin Heidelberg 2010

Colour composite of the Maltese Islands based on Landsat 7 ETM image, 9 January 1989

With a population of just over 31,000 inhabitants, Gozo is much less urbanized and greener than its sister island, Malta. One finds a more varied geology and larger relief contrasts, with typical flat-topped hills and fertile valleys. It boasts an impressive coastline with its sandy beaches, magnificent cliffs and unique karst features such as the Dwejra Inland Sea, the Azure Window and Calypso's Grotto.

Wied San Blas, one of Gozo's many picturesque valleys

Inside the Old Citadel

The history of Gozo goes back to prehistoric times. The first settlers are believed to have crossed from Sicily around the 5th millennium BC as witnessed by the oldest freestanding megalithic structures at Ġgantija and the Xagħra Stone Circle. Throughout the years, Gozo has been vastly influenced by the cultures and history of its various dominators including the Phoenicians, Romans, Arabs, Normans, Spanish, The Knights of St. John, the French and the British, who have all left their mark on the local culture and folklore. The Old Citadel is one of the most beautiful architectural complexes on the island. Upon the arrival of the Order of St. John in 1530, the Citadel was only a small, fortified medieval town. In 1551, the Turks invaded Gozo and besieged the "Castello". After a brief and heroic resistance the 5,000 inhabitants sheltering within the Citadel surrendered. With the castle destroyed, all those found within its walls were taken away into slavery and it took almost 50 years to repopulate the island and rebuild the Citadel to its present layout with its austere bastions. The Citadel now hosts various important buildings, most notably the Cathedral built during 1697–1703 by Lorenzo Gafa. The Cathedral is also the annual pilgrimage site of the Grand Priory of the Mediterranean of the Hospitaller Order of Saint Lazarus of Jerusalem.

The main source of income on this small island comes from agriculture, fishing and tourism. Traditional farming and animal husbandry practices are very widespread and the island is still home to several cottage industries, including lace making, glass blowing and local *ġbejniet* (goat cheese) making. Gozo welcomes thousands of tourists each year, including the hundreds that cross from the mainland to visit the island for some peace, or revelry. All these attributes make Gozo into a unique destination full of culture, history and charm.

Boat houses in Xlendi

Administratively, the island is divided into fourteen local councils. i.e. municipalities. The road network follows a star-shaped pattern with all roads connecting to Rabat (Victoria), Gozo's capital city. One of the most important archaeological sites, from the temple period in the Maltese Islands, the Ġgantija Temple Complex, is found in Xagħra Local Council, which also hosts Gozo's most famous beach with its beautiful golden-reddish sand Ramla l-Ħamra. The lighthouse at Ta' Ġurdan offers a 360 degree panorama experience of the island.

Geological Setting

> *"Northwest of the Gozo Great Fault the topography becomes increasingly dominated by the Globigerina Limestone series. The large dissected Upper Coralline Limestone plateaux of Nadur and Xagħra give way to the smaller remnant hills of Żebbuġ and Rabat, then to roughly conical hills like Ġiordan. Strong valleys between, which break abruptly through the Greensands in cliffs, give way to valleys cut almost entirely in the Globigerina rock. In the east, the karst tops are strongly dissected and further west broad valleys still cut through the complete sequence."*
>
> D.M. Lang, Soils of Malta and Gozo, 1960

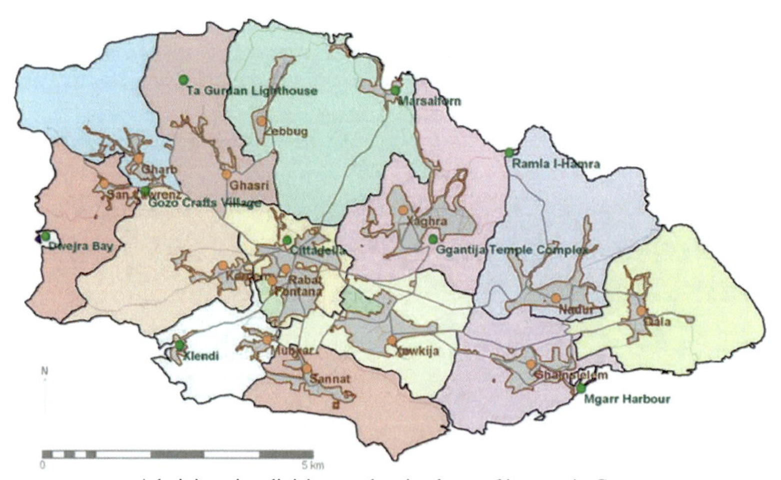

Administrative divisions and main places of interest in Gozo

Origin

The Mediterranean basin has been dated to the Late Triassic and Early Jurassic formed during the rifting of the African and Eurasian plates. The Maltese islands lie on what is called the Malta-Hyblean Platform that is found on the topmost margin of the African Plate. This sedimentary platform is the result of the accumulation of carbonate sediments deposited in a relatively shallow marine environment at a time when the eustatic level of the Mediterranean was higher than today. The Islands expose an Oligo-Miocene succession that is underlain by at least 3,000 m of late Oligocene to early Cretaceous (or Jurassic) shallow water carbonates.

The original strictly horizontal layers of the depositional platform were subjected to frequent "extensional tectonics" and subsequent folding, tilting, faulting, up- and over-thrusting. This finds the whole block of islands itself tilted eastwards, raising the cliffs to the west to about 240 m above sea level and drowning the valleys on the eastern and south–eastern coast. This tilting is also visible on the island of Gozo. From the south–west to the north–east, the coast consists entirely of cliffs, whilst the southern part of the island facing Malta is low lying.

Spectacular sheer cliffs mark the south-west to north-eastern coastline of Gozo

Geological Formations

The Maltese Islands are made up almost entirely of marine tertiary limestone with subsidiary clays and marls. All exposed rocks were deposited in shallow waters during the Oligocene and Miocene periods. Quaternary deposits represent the other type of rock formation found on the islands and are sediments that were deposited in a terrestrial environment following the emergence of the Maltese Islands above sea level.

The sequence of Tertiary rocks, starting from the uppermost and hence "most recent" formation, is shown in Table 1.

Table 1 Geological formations: thickness and lithological characteristics

Era	Formation	Thickness (m)	Lithology
Quaternary		2–10	Alluvial valley fillings, cave deposits
Miocene	Upper Coralline Limestone	30–100	Cross-bedded limestone with reef formation patches
	Greensand	0–12	Sand with phosphorite material
	Blue clay	0–75	Kaolinite rich clay and marly clay
	Globigerina	30–230	Pale yellow massive limestone beds
Oligocene	Lower Coralline Limestone	> 450	Coarse grained with extensive cross-bedding and reef patches

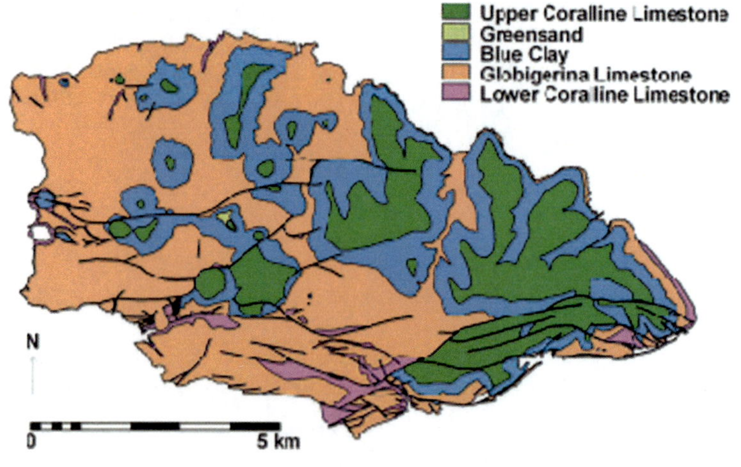

Geological Map of Gozo

Lower Coralline Limestone is the oldest exposed rock in the Maltese islands and is responsible for forming magnificent cliffs, some reaching 150 m in height. These

Lower Coralline Outcrop at Ta' Ċenċ Scutella bed in the Lower Coralline Limestone

characteristic vertical cliffs are found around most of Malta, south of Fomm ir-Riħ, and also in western Gozo. The Lower Coralline is hard and intractable and when found inland usually gives rise to a barren, grey limestone-pavement topography.

At the top of this formation is a characteristic band, a few metres thick, with frequent fossil occurrence, referred to as the *Scutella* bed. The widespread *Scutella* graveyard, found among other locations at the Dwejra Inland Sea, suggests that during the formation of the uppermost part of the Lower Coralline Limestone the sea floor was nearly flat.

Globigerina Limestone represents the second oldest rock and is the most widespread formation outcropping on the Maltese Islands. Softer than the underlying Lower Coralline Limestone, Globigerina produces rather meagre soils, and a gentle, rolling landscape. The latter characteristic has lent itself to terraced slopes which are intensively cultivated.

The Lower Globigerina provides the golden brown building stone, "franka", which is a very easily cut freestone which is worked in large, vertical sided quarries and is used for most of the buildings on the islands.

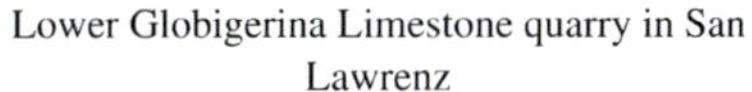

Lower Globigerina Limestone quarry in San Lawrenz

Blue Clay 'talus' in Ramla

Blue Clay is a series of blue and yellow clays and marls which overlie the Globigerina Limestone. Aided by the spring flows above it, Blue Clay forms the most fertile outcrop on the islands. The outcrop is usually narrow and the slopes steep, and is carefully terraced and tilled, except in places where it forms a classical "talus".

Blue Clay is of considerable importance to the fresh water resources on the Islands, as the formation acts as a seal to the water which infiltrates from the surface and is stored in the Upper Coralline Limestone and Greensand aquifers.

Greensand is a coarse orange-brown, thick-bedded fragmental limestone, which usually forms the base to the Upper Coralline Limestone cliffs. An abundance of glauconite gives the rock a green colour.

The formation has yielded a very rich echinoid fauna together with shark teeth and the remains of dugongs, manatees, dolphins and whales, indicating that the Blue Clay was deposited under warm shallow marine conditions.

Upper Coralline Plateau behind Ta' Pinu Church

Upper Coralline Limestone is the youngest Tertiary formation on the islands but resembles the oldest and lowest formation both in its chemical and palaeontological characteristics. Usually the Upper Coralline Limestone caps the highest hills which are typically barren limestone plateaux. In Gozo, these take the form of conical hills that are found in a uniform distribution across the island.

Pleistocene Quaternary Deposits are found in the form of palaeosols, fluvial gravels, coastal conglomerates and breccias, and bone deposits in caves and fissures.

Geological Faults: As in Malta, the island of Gozo is cut by a system of normal faults most of which follow a north eastern – south western (NE–SW) trend. A

Transverse faulting below Ta' Ċenċ

particularly dense sequence of such faults is found in the south eastern part of Gozo, around the Mġarr harbour. A second system of faults follows a north western – south eastern (NW–SE) trend and is particularly dense in the central, southern part of Goze. These geological faults play an important role in the replenishing of the aquifers.

Soils

The original, natural pattern of soils distribution in the Maltese Islands is largely determined by the parent material (in most cases the underlying rock), climate (both present and past under which rock weathering and soil formation took place), time and topography. The different kind of soils in Malta and Gozo are almost entirely due to the differences in the chemical composition, as well as in the physical constitution of the parent rock. The soils have in common that they are very calcareous because the parent rock is limestone. Since the processes of soil formation on calcareous material is very slow in a semi-humid country, the soils on the islands are considered to be young or immature soils.

The fertile Lunzjata valley

On the ridges, plateaux and plains, the soils are usually very shallow ranging in depth from less than 20 to about 60 cm. In valleys the soils are much deeper, often exceeding 150 cm, while on the gentler slopes of the Globigerina Limestone, the soils are moderately deep ranging from 60 to 120 cm.

The soils of Malta and Gozo were first classified by Lang (1960) following Kubiena's system into three groups: the Terra Rossa, Xerorendzina and Carbonate

Raw soils. It should be noted that today, the distribution of soils in Malta and Gozo appears to follow a rather random pattern. The Fertile Soil Act, particularly during the rapid urbanization witnessed throughout the 1980s, resulted in large quantities of soil being moved from one location to another as required for its preservation. Thus, the distribution of soils today no longer maintains the very close relationship to the underlying bedrock as at the time when Lang produced the Islands' Soil Map in 1960.

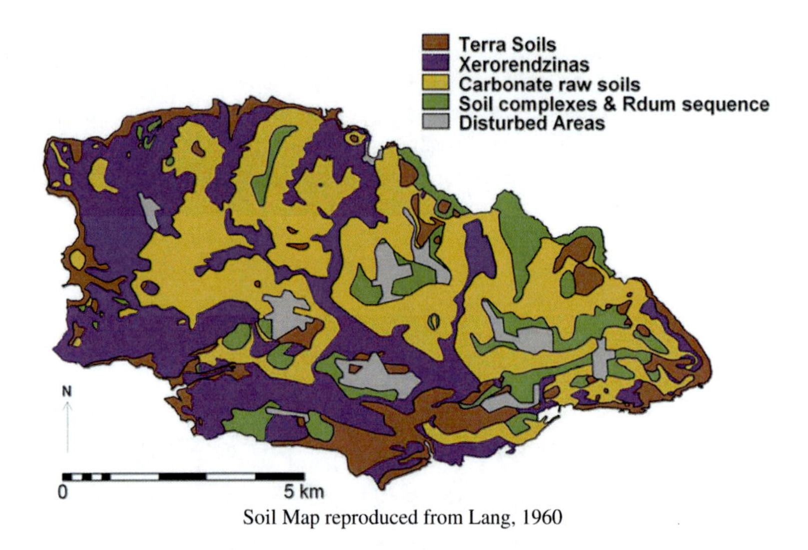

Soil Map reproduced from Lang, 1960

In recent years a national soil inventory of the soils of the Maltese Islands was undertaken within the framework of the project MALSIS, A *MAL*tese Soil *I*nformation *S*ystem. Information on soils is now found at the National Soil Unit (NSU), in a soils geo-database that includes soil classification information, and data on soil characteristics and soil contamination for more than 350 geo-referenced sites.

Water Resources

"Sir, I have inspected the Malta Water Works and
I find that they are in a highly satisfactory condition."

correspondence from Mr. Osbert Chadwick, dated 10th January 1896,
reporting to Count. G. Strickland, Chief Secretary to Government

A Brief History

In the absence of perennial streams, the scarcity of fresh water has been an acute problem for many centuries. This scarcity is more apparent during the summer period, when water availability has traditionally depended on the efficient collection and storage of rainfall during the rainy season.

Several tanks for the storage of fresh water have been discovered dating back as long ago as Punic and Roman times. The delegation of the Knights of St. John which was sent to report on the existing conditions on the Islands in 1530, described the water resources as being “salty and sedimentary”. The delegation observed that “the local population stored water in cisterns and even in ditches”. After the founding of Valletta by the Knights in 1566, several measures were taken to conserve water resources. The provision of fresh water was considered to be of paramount importance to the inhabitants of Valletta since the lack of it could have drastic consequences especially during a siege. Thus regulations were set up to prohibit gardens in the city and to enforce the construction of a well (*bir*) in every house. These measures were, however, insufficient and several attempts by various grandmasters followed to further secure the water supply in the city.

Between 1883 and 1897 engineer Osbert Chadwick devised several programmes which were aimed at addressing the islands’ chronic water supply problems. The loss of surface water runoff into the sea led Chadwick to propose that “small masonry dams be constructed at suitable points in all *widien* with an adequate catchment to appropriate the surface run-off during winter rains”. True to his life-long motto: “the first duty of an engineer is to make a measurement, his second to make a better one”, Chadwick went on to advise that “the ultimate height of each dam was to be determined experimentally, commencing with a low dam, and increasing its height according to the actual results observed from year to year” (Morris, 1952).

Water Catchments

The term *wied* refers to the valley (plural *widien*). The largest water catchment, Wied Marsalforn, includes the central part of Gozo which drains into the sea to the north. Table 2 lists the “larger” of around 40 distinct catchment areas that are found on the island in descending order of their surface area.

Table 2 Main water catchments in Gozo

Main catchments	Surface area (Km^2)	Main catchments	Surface area (Km^2)
Marsalforn	11.5	Mġarr	1.8
Ramla	6.3	Daħlet Qorrot	1.8
Xlendi	5.3	Wilġa	1.5
Mġarr ix-Xini	5	Biljun	1.1
Ilma	4.5	Raħeb	1
Mielaħ	3.9	Sabbar	1
Għasri	3.3	Pergla	0.9
Qliegħa	2.8	Xilep	0.3
San Blas	2		

Wied Marsalforn, Marsalforn Valley

Small dams are built to increase infiltration

From as far back as the 1880s, many small dams have been built across water-courses and are aimed at retarding the flow and thus to retain the water within the valleys *widien* for longer periods to allow increased infiltration and percolation into the limestone. The latter processes are responsible for recharging the aquifers from where the water either seeps out naturally or is extracted. Studies carried out by Morris (1952), Newbury (1968), and Chetcuti (1988) have consistently shown that only an estimated 16–25% of the annual rainfall percolates through the porous limestone rock to recharge the islands' aquifers.

Aquifers

Owing to the islands' geology, all aquifers are limestone aquifers. As in Malta, the two main categories of aquifers that are found in Gozo are the so-called "mean sea level" and "perched" aquifers respectively. As its description suggest, the former type of aquifer is found at sea level. Owing to its lesser density, the fresh water body within the rock formation finds itself "floating" on seawater. The Gheyben-Herzberg principle, which is based on the difference in density between fresh water and seawater, advises that for every one meter of fresh water above sea level, it can be expected to find up to 36 m of fresh water within the rock formation below sea level. The natural drainage of this aquifer occurs all around the coastline, and hence this aquifer takes the form of a lens-shaped body.

The Mean Sea Level Aquifer extends over the whole island with the exception of a small area in the south-eastern part of the island around the Mġarr harbour.

The perched aquifers on their part owe their existence to the presence of the Blue Clay formation. The latter acts as a seal to the deeper infiltration of water from the land surface. The overlying, porous and fissured Upper Coralline Limestone allows for substantial amounts of water to be retained in the Upper Coralline and (where present) Greensand aquifers. Thus, these aquifers are literally "perched" on the top of the Blue Clay.

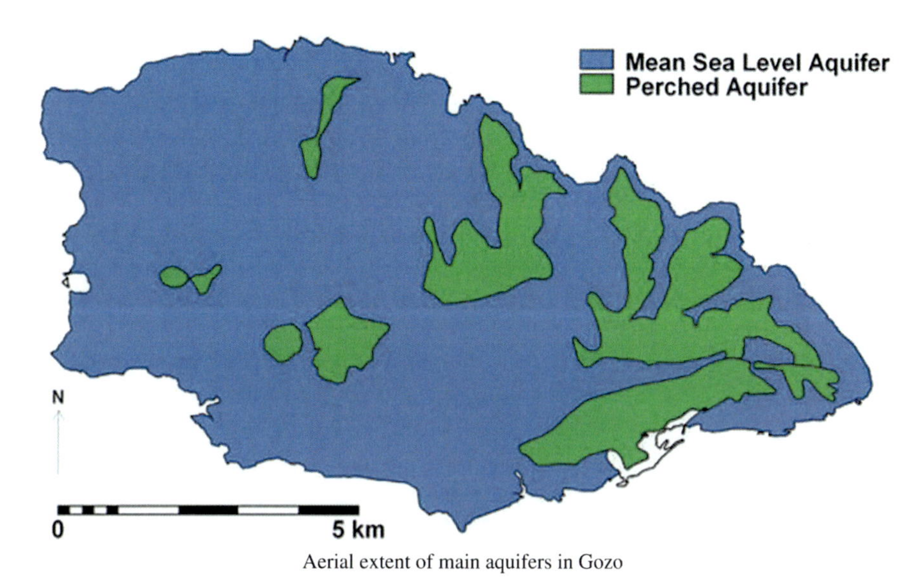

Aerial extent of main aquifers in Gozo

Table 3 provides an overview of the perched aquifers in Gozo, their aerial extent, the names of the major springs as well as the means of extraction and main use of the water.

Table 3 Overview of the Perched aquifers in Gozo

Perched aquifers	Surface area (km^2)	Major springs	Means of extraction[a]	Main use
Għajn Abdul	0.12	Għajn Abdul		Agriculture
Għajnsielem	2.88		224 private wells 1 borehole for public supply	Agriculture
Għar Ilma	0.10	Għar Ilma		Agriculture
Kerċem	0.29		old wells and a number of galleries	Domestic
Nadur	4.88		427 private wells and 37 springs	Agriculture
Victoria and Fontana	1.05	Għajn il-Kbira	old wells and a number of galleries	Domestic
Xagħra	3.02		475 private wells and 15 springs	Agriculture
Żebbuġ	0.38		82 old wells (*spejjer*)	Domestic

[a]Figures obtained from MRA (Malta Resources Authority), 2005, Initial Characterisation of Groundwater Bodies

Public Water Supply

With regard to the extraction of groundwater for public water supply purposes, Morris (1952) lists two main civil engineering works: "an older line of galleries, of total length 4,610 feet along *Wied Imġarr ix-Xini* (Mġarr ix-Xini Valley) on the southern side of the island, and a newer line, of total length, 6,020 feet, much of which was constructed during the war years, along *Wied ta" Marsalforn* (Marsalforn Valley) on the northern side of the Island'. These galleries are underground tunnels cut into the rock just above the mean sea level datum. The excavation proceeded from a vertical shaft; and the underground tunnel was then excavated away from this shaft while maintaining a slight, upward gradient. This technique assures that the water collected in the tunnel flows by gravity towards the shaft from where it is pumped to the surface. In Gozo, these underground galleries were dug along the natural course of the valleys.

The 1970s saw the advent of a large-scale programme of borehole drilling across the Maltese Islands. While the underground galleries supplied only what was being collected by gravity, the boreholes could (at least in theory) supply water at a constant rate all year round. In reality, this "new" method quickly led to increased levels of salinity in the water extracted. The mode of operation compromises the "natural equilibrium" between the fresh water body and the seawater below, thus leading to increased levels of sea-water intrusion in the Mean Sea Level Aquifer.

The National Statistics Office (2004) reported that extraction from boreholes in Gozo accounted for as much as 91% of the total water abstraction for drinking water supply purposes, with only 9% being produced by underground galleries. The Malta State of the Environment Report (SOER, 2005) confirms that major groundwater bodies in the Islands are being over-abstracted or are dangerously close to being over-abstracted.

Karst Heritage

"The importance of conserving representative karst areas
for science and recreation has been recognized in many countries
by the designation of national parks and reserves."

(Ford et al., 1989)

"Some of the best examples of normal faulting, karstification and solution subsidence, cliff recession, cave formation as a result of marine erosion,
and incision of steep-sided valleys to be found in the Maltese Islands occur here."

(Cassar et al, 2004) on the Qawra/Dwejra area in Gozo

Karst Landscapes

Karst landscapes or terrains represent a distinctive topography in which the landscape is largely shaped by the dissolving action of water on carbonate bedrock. This geological process, occurring over many thousands of years, results in dramatic landscapes with unusual surface and subsurface features ranging from sinkholes, dolines, vertical shafts, disappearing streams and springs, to complex underground drainage systems and caves.

Karst in the Maltese Islands

The Maltese Islands are characterised by well-developed karst phenomena which result in truly dramatic landforms. Several dolines as well as many caves (Calypso's cave in Gozo being the most legendary) can be found in the uppermost geological layer, the Upper Coralline limestone. The Lower Coralline hosts archaeologically important caves, sinkholes (il-Maqluba, Malta) as well as some spectacular natural arches (Azure Window in Gozo, Blue Grotto in Malta). Inland, this rock type forms barren, karstified limestone platforms (Ta' Ċenċ, Gozo). Far less karstified compared to the earlier limestone formations, karstic channeling and cavities in the Globigerina are found mostly where it outcrops.

Caves

The better known caves in Gozo include Calypso's Cave, Xerri's Grotto, Ninu's Cave and Għar Ilma. The former three caves are situated within the boundaries of the Xagħra Local Council. Calypso's Cave may not be considered an impressive cave in terms of karst features, but its association with Homer's epic "The Odyssey" always manages to generate curiosity among visitors to the island. Xerri's Grotto and Ninu's Cave, on the other hand, provide some remarkable natural speleothems, of both stalactite and stalagmite formations. In Xerri's Grotto there are also some interesting formations, which have developed as the result of calcification of tree roots. In Ninu's Cave, the calcification of water dripping from the cave ceiling formed numerous magnificent columns standing side by side. Għar Ilma provides an example of an important archaeological site due to the deposits that were found suggesting its use as a dwelling during Neolithic times.

The Gozo coastline is also dotted with a large number of sea caves. It is quite likely that some of these caves started as inland karst forms, i.e. underground caves which were later invaded by seawater. From this point onwards, the wave action became the dominant agent for the continued erosion. The mechanical action of seawaves and the receding cliffs have at times provoked the collapse of the ceiling of the caves of karstic origin.

An impressive example is the spectacular limestone archway found at Dwejra Point, referred to as the Azure Window. This is a natural arch rather than a window and was formed by the enlargement of an initial cave that developed along a line of rock weakness, which has cut through the limestone, resulting in the arch being

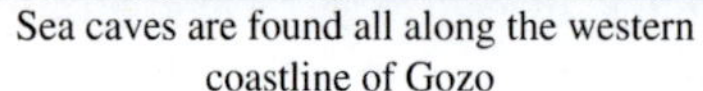
Sea caves are found all along the western coastline of Gozo

Azure Window at Dwejra

formed. The long ledge of rock forming the upper arch is in danger of collapsing and ultimately the roof will fall down and the westerly remnant will form a stack, similar to Fungus Rock.

Solution Subsidence Structures

Subsidence structure is the general term used for the different types of natural depressions occurring in karst environments, which are also referred to as dolines and sinkholes. The formation of these structures can be varied, but it is mainly associated with the solution of limestone by percolating acidified ground water which eventually leads to roof collapse of the enlarged underground caverns or caves. These structures can range from a few metres to a few hundreds of metres in diameter and in depth, and they may occur as isolated features (as for example the Maqluba sinkhole in Malta), or in groups as for example in the Qawra area in Gozo. Here, both the Inland Sea and Dwejra Bay provide excellent examples of large-scale circular subsidence structures that were formed during the Miocene period.

Dwejra Bay

Inland Sea

Fungus Rock or *il-Ġebla tal-Ġeneral* as it is locally known, is another karst feature of note in the Qawra area. It is found several metres away from the shore and is the surviving seaward edge of a collapsed subsidence structure. This outcrop, or stack, is surrounded by spectacular sheer vertical cliffs and is home to the rare phallic-shaped Fungus Melitensis, *(cynomorium coccineum)*. Despite its many given names referring to it as a fungus, the plant is neither a fungus nor is it found only in Malta or Gozo. The Knights of St. John believed this plant to possess medicinal powers and went to great extremes to guard the Fungus Rock and protect the much-prized plant. Anyone caught stealing the plant was sentenced to death or put to the galleys.

Fungus Rock

Other Karst Features

Other features in the karst landscape of Gozo include grikes and clints, as well as a dense development of solution pits and solution pans.

Grike and clint topographies are common in karst landscapes and are an assemblage of irregular, deep, narrow grooves present at the rock surface, also known as "limestone pavement". These forms are referred to as "Karren" (German) or "Lapies" (French).

Solution pits are round-bottomed erosion holes that are usually circular, elliptical or irregular. Solution pans, on the other hand, display a flat or nearly flat bottom that is usually horizontal. In the past, these solution pans were sculpted into a more regular arrangement by man and used as salt pans. An example of such salt pans, which are still in use to date, can be seen at Marsalforn.

Typical grike and clint topography

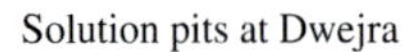

Solution pits at Dwejra

Salt pans at Marsalforn

The Special Vulnerability of Karst Environments

All over the world, karst landscapes are arguably among the most vulnerable environments because of the close interconnectedness between the geological, geomorphological, biological, and most importantly, hydrological processes affecting these landscapes. The underground solution channels and conduits commonly provide a direct hydraulic link between the surface and the underground aquifers. This causes the hydraulic behaviour of karst areas to be very complex. This can be illustrated by the fact that it is common in karst environments for the impact of an accidental spillage to be transmitted through underground path flows to a point which may be far removed from the initial point of impact. This is a distinctive characteristic of a karst environment, with its hydrological processes that operate underground and, unfortunately, out of sight of many policy makers (Urich, 2002). Quite literally, a karstified drainage system may lead to the development of blind or disappearing streams. A typical example of a blind valley is found at Ħondoq ir-Rummien, in the south-eastern part of Gozo.

Blind valley at Ħondoq ir-Rummien

Production of a Water Resources Vulnerability Map for Gozo

The production of a water resources vulnerability map for the island of Gozo featured as one of the innovative outputs achieved from our research carried out under the EU-sponsored International Cooperation (INCO) Project entitled *ResManMed:* Resource Management in Karstic Areas of the Coastal Regions of the Mediterranean The SCI (Surficial Cover Infiltration) map provides a qualitative assessment of the relative ability for contaminants at the surface to infiltrate down to the underlying aquifer.

The methodology which is used to produce the SCI map considers the combined appraisal of four hydrogeological factors: surface lithology, faults, karst features and surface drainage density. Developed by the authors in partnership with the International Research and Application Centre for Karst Water Resources at Hacettepe University in Ankara, Turkey and the National Centre for Remote Sensing in Beirut, Lebanon, the methodology consists of first rating the effect of each of the hydrological factors individually. Subsequently, by assigning weighting coefficients to the factors, their combined effect is represented by the so-called SCI Index. The latter index is calculated for each cell of a grid that has been placed over the area under investigation. For the Island of Gozo, the SCI Indices have been calculated according a grid made up of 200 by 200 m^2 elements.

Calculation of the SCI Index

In each cell of the grid, the SCI Index is calculated on the basis of the ratings obtained for each of the individual factors. The four factors have been assigned with a "weight" which represents their relative importance on the infiltration phenomenon. (Table 4).

Table 4 Relative weights assigned to the factors considered in the SCI Method information on the Perched Aquifers in Gozo

Factor	Weight (%)
1- Surface lithology	35
2- Faults	20
3- Karst features	30
4- Surface drainage density	15

The SCI Index in any given cell of the grid is obtained as follows:

$$\begin{aligned} \mathit{SCI\ Index} &= 0.35x\ (\mathit{Surface\ Lithology}) + 0.20x\ (\mathit{Faults\ rating}) \\ &\quad + 0.30x\ (\mathit{Karst\ Features\ rating}) + 0.15x\ (\mathit{Surface\ Drainage\ Density\ rating}) \end{aligned}$$

Finally, the SCI Index values are interpreted according to the colour legend shown below. (Table 5).

Table 5 Classification and colour legend of SCI Index values

SCI Index value	Colour code	Interpretation
<0.40		No infiltration
0.41–0.85		Very low infiltration ability
0.86–1.30		Low infilration ability
1.31–1.75		Moderate infilration ability
1.76–2.20		High infilration ability
> 2.20		Very high infitration ability

In the maps provided below, the effect of the hydrogeological factors on the final SCI map is examined in more detail.

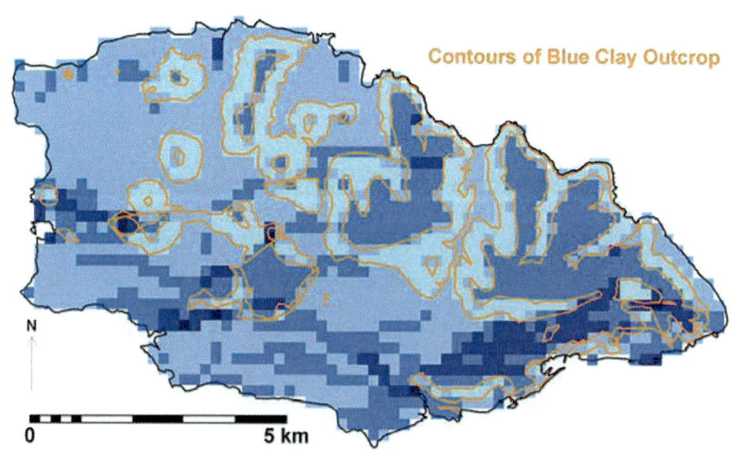

Effect of Blue Clay Outcrops

The SCI method considers that the surface lithology is divided into three distinct classes: pervious, semipervious and impervious, which are assigned with a rating of 2, 1 and 0. The infiltration ability is reduced in those areas where the Blue Clay forms the outcropping formation (delineated by the contours shown in orange). These areas are shown to have a "Very low infiltration ability" in the SCI map.

Fault lines play an important role in the infiltration process. The SCI method measures the length of the faults in each cell of the grid, leading to a fault density value. The density values thus obtained are divided into 3 categories: high, medium and low and are assigned with ratings of 3, 2 and 1. Areas with a high fault density are clearly increasing the infiltration ability on the SCI map.

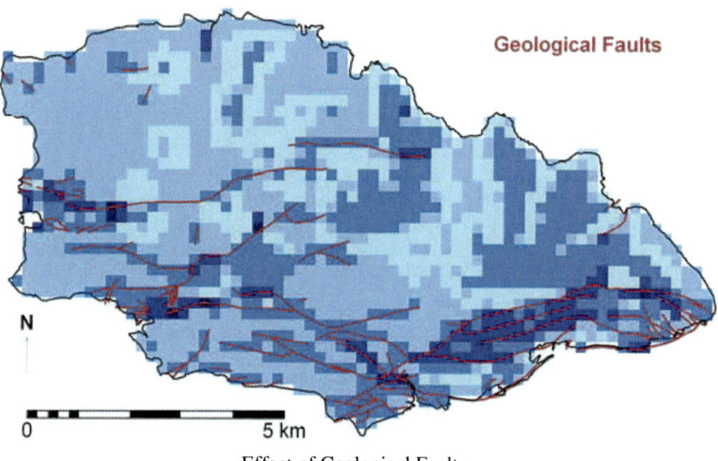

Effect of Geological Faults

According to the SCI method, the cells that contain depression features of karstic origin are assigned with a higher infiltration rating. Their surface area is measured in each cell of the grid and assigned with a rating of 3 (high), 2 (medium) or 1 (low density). However, in areas where the surface lithology is impervious, a 0 rating is assigned, i.e. in this instance the SCI method considers that the karst feature does not have any effect on the infiltration.

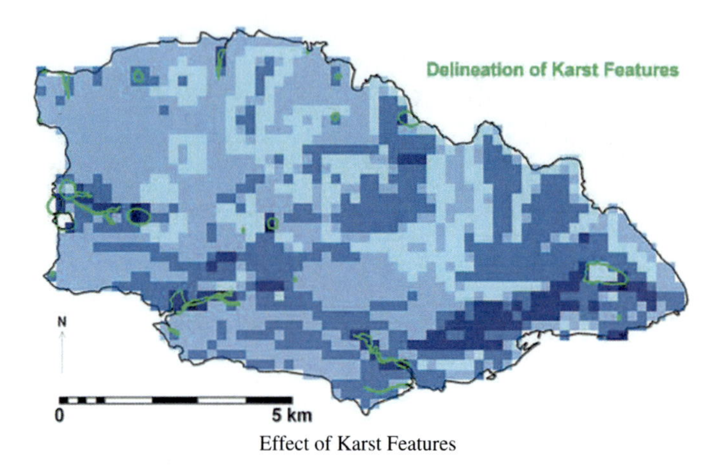

Effect of Karst Features

One of the largest dolines is found in the south-eastern part of the island however, its location coincides with a Blue Clay outcrop. The latter represents an impervious surface lithology. Thus, in agreement with the method, this particular doline is considered to have no effect on the infiltration ability. Apart from dolines, the karst features which affect the relative infiltration ability for the island of Gozo include a series of gorge shaped valleys, which appear as deep incisions into the landscape.

Wied ix-Xini, a gorge shaped valley

The final, fourth factor considered in the SCI method is the surface drainage density. The study of the drainage (stream) network on a topographical map can provide useful hints about the hydrogeological conditions. For example, areas that show a very dense drainage network can typically be associated with the existence of impervious formations beneath the surface cover. Thus, a higher Surface Drainage Density is linked to lower infiltration ability.

In Gozo, the Surface Drainage Network is quite uniformly distributed across the whole island. Moreover, the Surface Drainage Density values are consistently "low". This led to the understanding that the Surface Drainage Density factor should not be used to further "fine tune" the SCI Index. The SCI map considers the Surface Drainage Density to be uniform and low, and thus with a maximum rating of 3 for the whole island.

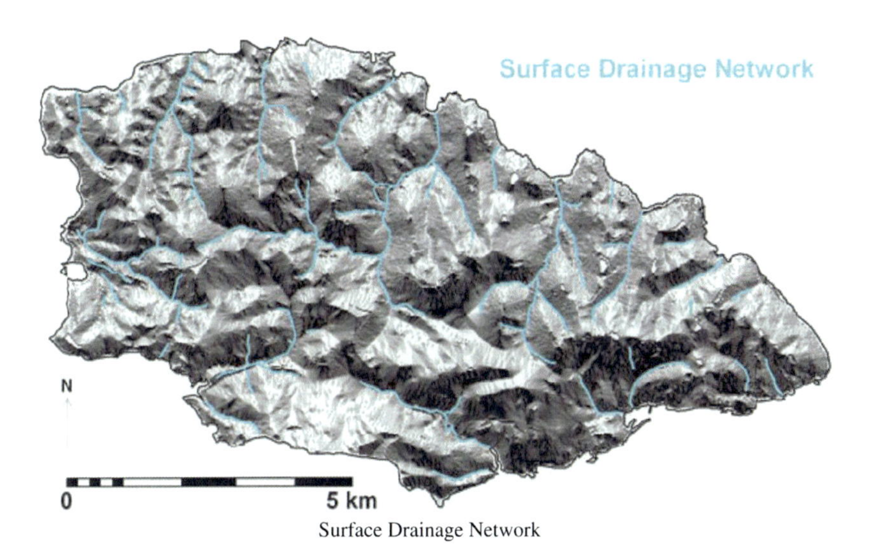

Surface Drainage Network

Human Impact

"For over 5,000 years people have lived here, and have changed and shaped the land, the wild plants and animals, the crops and the constructions and buildings on it. All that speaks of the past and the traditions of the Islands, of the natural world too, is heritage."
Haslam and Borg (2002)

From an infrastructural and urban development point of view, Gozo is generally still less developed than the main island, Malta. However, its environmental resources are acknowledged to be under constantly increasing pressure from human activities. Urban development, quarrying, waste disposal, overexploitation of groundwater, down slope ploughing and arable land abandonment have intensified the pressures on the island's natural resources.

Limestone Quarrying

The increased urbanization in Gozo over the past few decades has intensified quarrying practices on the island, wearing away stretches of the already limited natural and agricultural land. Quarries are mostly concentrated in the north western part of Gozo, mainly in San Lawrenz, but also in Kerċem and Għarb Local Councils. The majority are so-called soft stone quarries, quarrying the Globigerina Limestone, the lower part of which produces the highly sought after building stone called "franka".

The environmental impact associated with the quarrying of limestone needs to be analysed carefully in relation to the protection and conservation of the ground water resources. Indeed, the quarrying of limestone represents the removal of the natural "protective cover" overlying the aquifers. Abandoned quarries often become utilized as waste sites. Unless the necessary precautions are taken, this practice can dramatically increase the likelihood of pollution.

Quarrying at Dwejra

Groundwater Exploitation and Pollution

According to the Malta State of the Environment Report 2005, a "preliminary risk assessment carried out by the Malta Resources Authority indicates that, with the exception of the Comino Mean Sea Level aquifer system, all Malta's groundwater bodies are at risk or probably at risk of failing to meet the objectives of the (European) Water Framework Directive: 'Malta's ground waters are seriously at risk from overexploitation and pollution, risking the loss of Malta's only renewable freshwater resource".

The highly fissured limestone makes the possibility of sea water intrusion into the Mean Sea Level aquifer a permanent problem, yet the over-abstraction of the aquifer has clearly exacerbated the problem to an alarming level.

The dumping of waste is a major source of concern. Derek Ford (1989) alerts us, saying "Unfortunately, in all inhabited karsts (around the world), dolines and sinkholes are perceived as being particularly suited for the dumping of solid or liquid waste, because it disappears underground and 'out of sight is out of mind!' As elsewhere in the Mediterranean, besides dolines and sinkholes, it is often whole valleys that are the 'preferred' target for the illegal dumping of waste!"

Evidently without adequate and rigorously enforced protection measures even regulated landfills present a major problem. A heritage conservation area, which appears to be under particular threat, is found in Dwejra. This major tourist attraction is in close proximity to a landfill, which is growing at an alarming rate. Besides damaging the aesthetic beauty of the area, it also threatens to pollute the marine environment from leachates reaching the sea.

As has already been mentioned, a major threat to the quality of ground water derives from quarrying activity and, even worse, from land filling when a disused quarry is not strictly monitored to receive inert waste only.

Agricultural Practices

Agriculture has always been an important pillar of Gozo's agrarian economy. Over time, in an effort to conserve and exploit the barely adequate soil and water resources, human activity has literally re-sculptured the land surface by cutting terraces through the soft rock, building stone embankments and walls, and redistributing the stones and soil material on the terraces. Traditionally, the slopes of valleys are terraced to allow agricultural cultivation. Every cultivable piece of land with soil has been terraced and is used to grow some crop at some time or other.

Agriculture has had a great effect in shaping the island's environment by providing such characteristic landscape elements as rubble walls, terraced fields and wind pumps. However, increasing intensification of agriculture can also have a negative impact on the quality of the landscape.

Traditional farming

Today because of a high level of economic development, off-farm resources such as feed, fertilizers, pesticides, fuel and machinery are being used as substitutes for traditional farm inputs. From a resource-management point of view, despite this increase in mechanisation, management of arable land still has its shortfalls. Maintenance of the traditional rubble walls appears to be a dying practice and tillage of fields is not always done in parallel to the contours, thus facilitating erosion.

References

Cassar, L.F. et al. 2004. Report on a Survey of the Terrestrial Ecological Resources of the Qawra/Dwejra Area, Western Gozo, commissioned by Nature Trust (Malta).

Ford, D. & Williams, P. 1989. Karst Geomorphology and Hydrology, Unwin Hyman Ltd., London.

Haslam, S.M. & Borg, J. 2002. Let's Go and Look After Our Nature, Our Heritage!, Ministry of Agriculture & Fisheries – Socjeta Agraria, Malta.

Lang, D.M. 1960. Soils of Malta and Gozo in Colonial Research Studies No. 29, Colonial Office, London.

MALSIS (Maltese Soil Information System) 2004. Soil geographic database of the Maltese Islands. National Soil Unit, Ministry for Rural Affairs and the Environment, Malta.

Morris, T.O. 1952. The Water Supply Resources of Malta. The Government of Malta, Malta.

MRA (Malta Resources Authority) 2005. Initial Characterisation of Groundwater Bodies within the Maltese Water Catchment District under the Water Policy Framework Regulations, 2004. MRA, Malta.

ResManMed, unpubl. Interim and Final Reports on the INCO-DC project ERBIC18CT970151, Resource Management in Karstic Areas of Coastal Regions of the Mediterranean, 1997–2000.

Urich, P.B. 2002. Land use in karst terrain: review of impacts of primary activities on temperate karst ecosystems. Science for Conservation 198, New Zealand Department of Conservation.

The Geomorphological Cave Features of Għar il-Friefet

Anna Spiteri, Michael Sinreich, and Dirk De Ketelaere

Introduction

Għar il-Friefet is located in the south-eastern part of Malta, on the bank of a dry valley and in close proximity to the well-known cave named Għar Dalam. In July 2003, the Malta Environment and Planning Authority commissioned an ecological and a geological survey of the cave in response to an application for a new residential development in the cave's near vicinity. The main objective of the latter survey involved an assessment of the structural integrity of the cave. Given that there is no public access to the cave, the survey offered the authors a unique opportunity to record the cave's geomorphological features.

View of Wied Dalam from cave entrance

Għar il-Friefet is a geomorphological phenomenon as a whole that evolved through what is known as karstification, i.e. the dissolution of carbonate rock. In this instance, the karstification process resulted in the formation of a cavity with a

A. Spiteri (✉)
Integrated Resources Management (IRM) Co. Ltd. Malta
e-mail: info@environmalta.com
website: www.environmentalmalta.com

N. Evelpidou et al. (eds.), *Natural Heritage from East to West*,
DOI 10.1007/978-3-642-01577-9_30, © Springer-Verlag Berlin Heidelberg 2010

volume of about 2200 m^3 within the limestones of the Lower Coralline formation. The cave displays specific, interior geomorphological features which contribute to the understanding of how the cave, during its multi-phase evolution, came to develop into its present shape. There is also evidence of past and present fluvial geomorphological processes at work at the land surface especially in the Wied Dalam valley, outside the cave's entrance, which may project a continuation of the Għar il-Friefet cave system across the valley.

Location and Extent of the Cave

The cave consists of two levels, an upper and a lower level, which are shown as: "Level I" "Level II" respectively. Both levels are found within the Lower Coralline Limestone, which outcrops along the bottom of the valley called Wied Dalam. This

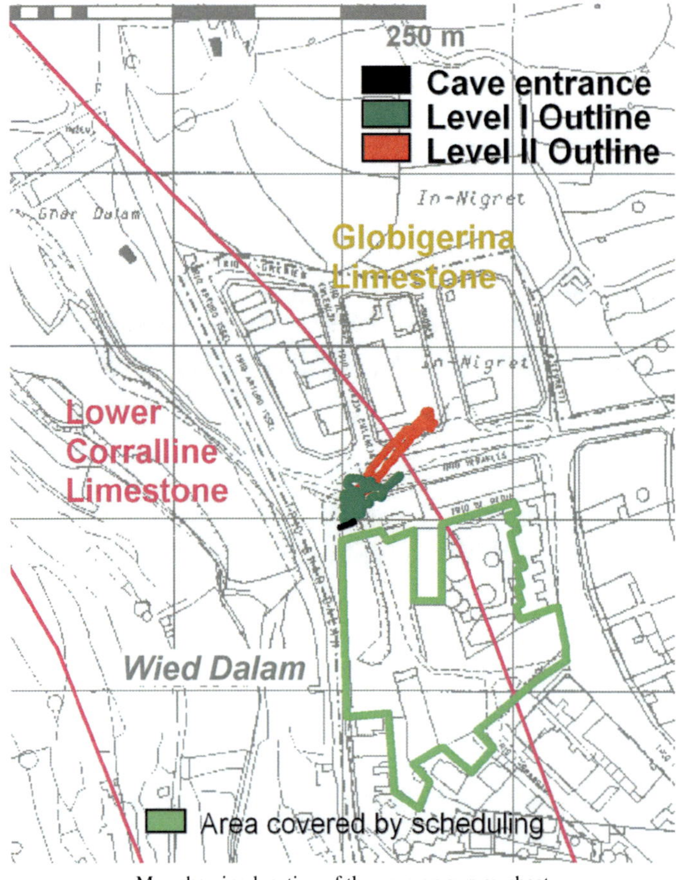

Map showing location of the cave on survey sheet

formation is overlain by the Globigerina Limestone, which outcrops above the latter part of the lower level. The area covered by scheduling, i.e. the area which was deemed to "protect" the cave from development, proved to have been defined outside the actual location of the cave. Instead, the entire cave complex has been covered by residential development and roads.

Cave Geometry

Level I, the upper level consists of one long chamber which has been sub-divided in Sects. 1, 2, and 3 for easy reference. The chamber is characterised by several elliptical voids (phreatic pockets), located to the west of the chamber, and ends with an intricate solution conduit made up of several shafts. The latter shafts have a diameter ranging from 35 to 50 cm. A shaft near the entrance of this solution conduit connects to yet another solution conduit which rises practically vertically from the main chamber and ends in the basement of a residence. Stone walls and arches, built to support the residences above, show the human intervention inside the cave.

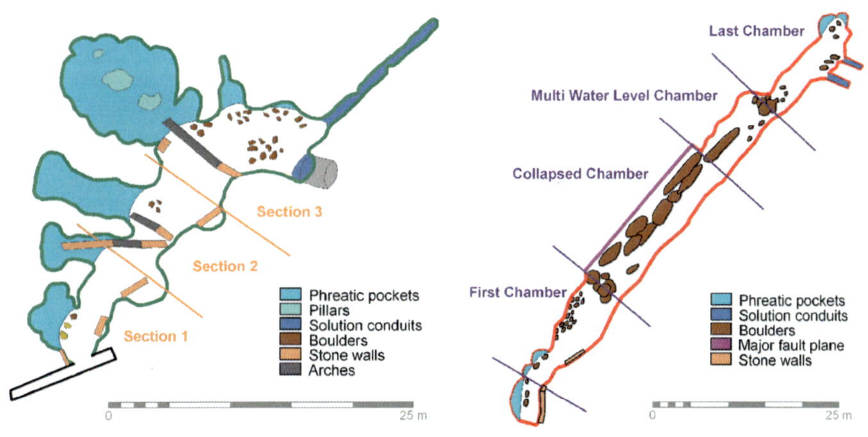

Plan view of Level I, the upper level

Plan view of Level II, the lower level

Level II has a distinct linear shape as it is to a large extent structurally controlled by a fault plane, exposed over a distance of more than 20 m, and with an average height of 8 m. For easy reference, Level II has been divided into different chambers, starting with the "First Chamber". This is followed by the "Collapsed Chamber" which is obvious from the high number of large to very large boulders that cover this part of the cave. This chamber is delineated by the exposed fault face. Then comes the "Multi Water Level Chamber", so called because it has several very clear cut paleo water table notches on one side. Finally, yet also the most difficult part of the cave to access as it is necessary to crawl and squeeze for almost 5 m in wet soil, the "Last Chamber" is reached.

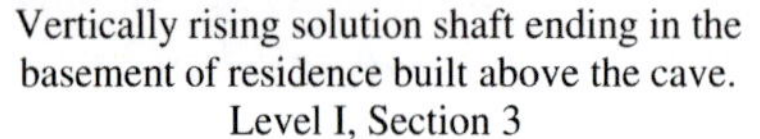
Vertically rising solution shaft ending in the basement of residence built above the cave. Level I, Section 3

Difficult access to the Last Chamber, Level II

Geomorphological Features

Phreatic pockets. Għar il-Friefet displays several classic examples of phreatic elliptical pockets, both in Level I and in Level II. In several cross sections, the cave profile has an elliptical shape indicating phreatic (pressure flow) hydrogeological conditions during cave formation. The elliptical shape of these pockets, with an elongated horizontal diameter is attributed to solutional attack at the weakest points of the bedding plane or fracture zone. These elliptical phreatic pockets were formed when the cave was situated below groundwater level and the cavities were filled

Natural pillar inside phreatic pocket, Level I, Section 3

with water. Under these saturated conditions water flow caused both chemical corrosion of the carbonates and mechanical erosion of the classic parts of the lithology. Saturated water flow occurred in a direction south–west towards the today's cave entrance. It can be noted that the predominant distribution of elliptical phreatic pockets in the cave are found on the left side of the cave system, both in Level I, Sect. 1, 2 and 3 and in the Last Chamber of Level II.

Solution conduits. Solution conduits are found in both levels of the cave, and were formed during the phreatic phase of the cave, as is proved by their elliptical shape. These steep conduits acted as hydraulic connections between different cave levels. Tectonic movements allowed karstification to proceed downwards with the lowering groundwater table by developing deeper cavities. Once the upper part of the system, Level I, became dry and lost its phreatic regime it changed from saturated lateral water flow into an unsaturated vertical water percolation and/or stream flow respectively. In such vadose conditions solution conduits act as preferential percolation pathways.

Entrance to solution conduit, Level I, Section 3

An intricate solution conduit with several vertical shafts is found at the far end of Level I, Sect. 3. Once inside this solution conduit, it was noted that the furthest shaft was blocked by boulders, (either man-filled when the housing development occurred above or boulders collapsed inadvertently into this shaft during excavation). The cave's continuation in this direction has already been removed by surface erosion processes.

In the Last Chamber of Level II, a thick layer of wet sediments originates from the two conduits that are found here. Due to the very steep inclination and slippery conditions, it was difficult to explore the conduit system of this Last Chamber in more detail. Whilst most elliptical pockets are on the left of the cave system, the solution conduits are located on the right of the cave. The solution conduits may have been the original source of water brought to the cave.

Pendant in Last Chamber, Level II

Pendants. Għar il-Friefet also boasts of two classic examples of pendants found in Level II, respectively in the middle and on the right of the Last Chamber. Pendants are residual pillars of rock between anastomosing channels. There are various ways of how they can develop but in this case these gently rounded, hanging pendants were carved by fast flowing water draining down on both sides of it. The presence of these rounded pendants denotes a phase where groundwater must have been flowing through the subsurface via separate conduits which were later on merged due to progressing karstification by forming the cave cavity. They are thus a demonstration of the evolution of the cave with time.

Corrosion notches. Corrosion notches are found in Level II on the left of the First Chamber and again on the left side of the Last Chamber. These are commonly formed where, in a standing pool, water convection carries fresh water to the rock walls at the water surface. A sharp notch is dissolved there, tapering off very steeply below the waterline. These features are indicators for partly vadose conditions in the cave. The distinct notches found in Level II also suggest mixing corrosion at the interface of the karst groundwater and seawater which may have occurred before the uplift of both chambers or seawater level lowering respectively.

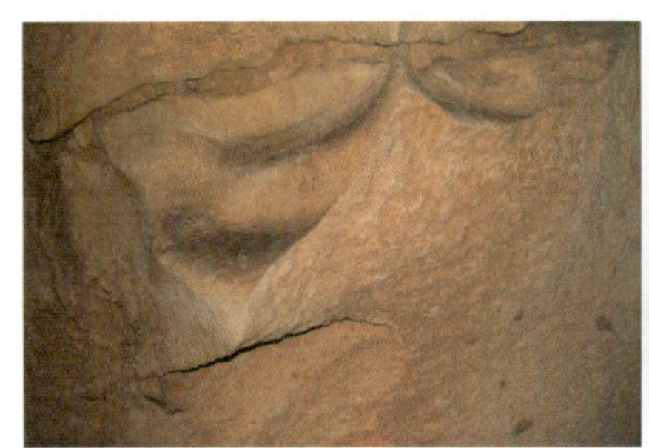

Corrosion notch, First Chamber, Level II

Multi Water Level Chamber, Level II

In the Multi Water Level Chamber, there are several corrosional notches that signify paleo-water table levels one above the other very precisely. Also in the Last Chamber there is evidence of such notches on the left side of the chamber. It is worthy of note that during the Quaternary, sea level oscillation occurred in the range of +5 to –80 m, over comparatively brief time spans.

Speleothems. Speleothems are carbonate dissolution on the one hand and precipitation on the other hand, which is still an ongoing process. Rainwater infiltrating in the subsurface above the cave percolates through the soil and the fissures of the uppermost part of the limestone. In this way, carbonates are dissolved and transported downwards. When reaching the cave ceiling, the carbonate precipitates by forming speleothems, like small calcite tubes. Particularly the phreatic pockets display an array of speleothems hanging from their ceilings. Calcite ligaments or flowstone, which is a deposit formed by thin films or a trickle of water over walls or floors, are found on both sides of the cave in Level I Sect. 1. Speleothems give clear evidence of vadose conditions during their formation. However, major speleothems, such as stalactites and stalagmites, are not present in the cave. This is due to dry climatic conditions since the vadose cave regime has established with limited percolation water availability.

Flowstone, Level I, Section I

Fluvial deposits. Fluvial deposits are mainly found on the left of Level I Sect. 1 and on the left of the First Chamber in Level II. During vadose conditions and perhaps after the uplift of the chamber occurred, a stream must have flowed through, bringing with it a bed load of suspended solid matter and pebbles. On the left of these chambers the stream channel has undercut the cave walls and thus encouraged roof collapse so that it is difficult to separate what is the former bed load and what is the result of roof collapse. However, the semi-angular shape of the rocks gives rise to the conclusion that most of them originate from the ceiling above and that they must have been carried for only a short distance and possibly only during intermittent periods of flooding, if at all. In the Last Chamber of Level II a thick layer of recent fluvial deposits consisting of clayey material is found. This may originate either from surface material swamped into the cave, or, which is probably the main mechanism, as residuals from clay-rich strata during karstification.

Recent, thick, layer of fluvial deposits, Level II, Last Chamber

Breakdown. In practically every part of the cave there is evidence of roof collapse. This is apparent from the jagged surfaces of rupture in the walls and in the roofs of the cave and the angular shape of the piles of rocks found. These shapes are consistent with the cuttings of tectonic joints and ruptures in the carbonate rock. Mechanical failure within or between rock beds and weak joints are normally attributed to breakdown in most caves leading to further cave genesis. However, several processes may explain roof breakdown experienced at Għar il-Friefet.

Fault plane (left) and massive monoclinal block, estimated at 60 m^3 (right), Collapsed Chamber, Level II

The uplift and water level decline respectively would have drained the cave system and removed the buoyant force of water, giving rise to mechanical failure. Additionally, stream erosion during subsequent vadose conditions, and the ongoing karstification by seepage waters would also have led to the weakening of the roof. Tectonic processes would also have caused the major falls that are found at the very end of the First Chamber, throughout the Collapsed Chamber and at the entrance to the Last Chamber.

Fault breccia, First Chamber, Level II

It is not difficult to distinguish between boulders resulting from roof collapse through mechanical failure and the fault breccia caused by major tectonic movements. Two typical samples of the rocks of the former type measure 44 cm × 54 cm × 80 cm and 35 cm × 54 cm × 70 cm, resulting in a volume of 0.19 m^3 and 0.13 m^3 respectively. The boulders caused by major tectonic movements are on average much larger. Two typical samples measure 1.50 m × 2.00 m × 2.50 m and 1.70 m × 1.40 m × 1.40 m respectively, corresponding to a volume of 7.50 m^3 and 3.33 m^3 respectively. The boulders of the Collapsed Chamber are again of a much higher magnitude, with one particular monoclinal block estimated at 60 m^3.

Further Reading

De Ketelaere, D. & Spiteri, A. 2004. Malta and Gozo Islands (Malta) in a contribution by members of the COST-621 Action Groundwater Management of Coastal Karstic Aquifers, The Main Coastal Karstic Aquifers of Southern Europe, edited by Calaforra, J.M., 2004. European Commission, Directorate-General for Research, Luxembourg.

Ford, D.C. 2003. Perspectives in karst hydrogeology and cavern genesis. – Speleogenesis and Evolution of Karst Aquifers 1(1), 12 pp found at www.speleogenesis.info.

Ford, D.C. & Williams, P. 2007. Karst Hydrogeology and Geomorphology. John Wiley & Sons, Chichester.

Morana, M. 1987. The prehistoric cave of Għar Dalam. Printwell Ltd., Malta.

Paskoff, R. & Sanlaville, P. 1978. Observations géomorphologiques sur les côtes de l'archipel maltais in Zeitschrift für Geomorphologie, 22/3, 310–328.

Shaw, T.R. 1952. The caves of Malta in Bulletin of the National Speleological Society, 14, 34–41, Washington, D.C.

Our Ancestral Country Allies: The Rubble Walls

Josianne Vella and Jesús Garrido

> *"rubble walls (are hereby declared) as protected, in view of their historical and architectural importance, their exceptional beauty, their affording a habitat for flora and fauna, and their vital importance in the conservation of the soil and of water"*
> The Rubble Walls and Rural Structures (Conservation and Maintenance) Regulations Legal Notice 160 of 1997, Malta

The Story of the Maltese Rubble Walls

Rubble Walls, locally known as *Ħitan tas-Sejjieħ*, are a predominant and integral feature of the Maltese rural landscape. These traditional agricultural structures reflect the history, knowledge, and skill of our ancestral agrarian societies. For ages, Maltese farmers have realized the important role that these walls play in the preservation and sustainability of the local agricultural economy.

Terraced fields with rubble walls – the Maltese rural landscape (Wied Għemieri, Malta)

J. Vella (✉)
Integrated Resources Management (IRM) Co. Ltd. Malta
e-mail: info@environmalta.com
website: www.environmentalmalta.com

N. Evelpidou et al. (eds.), *Natural Heritage from East to West*,
DOI 10.1007/978-3-642-01577-9_31,

Lack of good quality soil or water can make agriculture a hard and unprofitable business. The limited surface area, the hilly topography, the water scarcity and the meagre soils of these islands have always been a challenge for the local agricultural sector. In an effort to exploit the limited resources available, farmers have literally re-sculptured the land surface by cutting terraces across the hilly landscape. The use of terraced fields allows for sloping areas to be cultivated and is also designed as a means to slow surface runoff and prevent the soil from being washed away. In Malta and Gozo, terracing supported by rubble walls has been used by the local farmers for many centuries.

Rubble walls are dry stone walls, that is, walls entirely built without the use of cement or mortar. Their stability comes from the skilful placing and fitting together of the stones. The use of basic dry stone building can be traced back to prehistoric times and examples of dry stone walls and buildings can be found all over the world throughout history. In the Maltese Islands, early examples of dry stone walls can be seen around the entrances of the Megalithic Temples, dating back as far as 5000 BC, and also at the Bronze Age village of Borg in-Nadur.

Rubble walls are dry stone walls (Wied Qirda, Malta)

It was however, during the Arab occupation (870–1127 A.D.) that the construction of rubble walls became a widespread agricultural practice. The Arabs established important agricultural practices and introduced new irrigation techniques and also new crops. Rubble walls were used not only to delineate the boundaries and ownership of agricultural land, but also as a measure against soil erosion in terraced fields. Ever since, this network of rubble walls surrounding agricultural fields has dominated the Maltese rural landscape.

Transforming the natural landscape can have severe consequences. In fact, throughout history, human intervention has continuously changed and manipulated the environment in favour of new developments that have often been detrimental.

Rubble walls however offer a sustainable land use practice that supports agricultural activity in an environmentally unobtrusive way. The practice and use of terracing and rubble walls represents an approach to landscape transformation that is based on a long-earned understanding of the interrelationship between land resources and human activity. Land transformation for agriculture is usually necessary and indeed, at times vital for the survival and sustenance of communities and the population at large. Land transformation practices which respect and sustain the natural environment on which agriculture depends should be well preserved and appreciated.

Rubble walls as silent warriors (Wied Qirda, Malta)

In this respect, the knowledge contained within the traditional practice of rubble walls is a real gift. When thinking about the future, we must not disregard or forget traditional practices. Most people accept these rubble walls as integral elements of our local rural character, however, few realise their extreme importance. Rubble walls are like *silent warriors* who protect our livelihood and our quality of life in the tranquil setting of the Maltese countryside.

A Maltese Building Technique

The use of dry stone walls as field enclosures has been practiced in many countries throughout the centuries, but the type, and method of construction of these walls varies according to the region. In general, the type of wall built depends on the topography and natural resources available in the surrounding area. As has been the case in many ancient cultures around the world, the old agrarian communities in Malta developed their local terracing and wall building techniques using locally available materials. Consequently, the local rubble wall, *Ħajt tas-Sejjieħ*, is a unique feature of our rural landscape, part of our heritage and an important element of our identity.

Maltese rubble walls make use of natural, unhewn stones of different sizes, which are found in the countryside, hence the term "rubble". These stones are called *ġebel tax-xagħri* and are usually made of the hard-wearing upper coralline limestone.

In her book, *Ħitan ta-Sejjieħ*, architect Elizabeth Ellul (2005) provides a detailed study of the construction methods, the tools employed and the different types of rubble walls found on the islands. She identified three basic rules in the building of these walls:

1. the largest stones are placed at the base of the wall, except those used as tie-stones
2. there should be an infill in between the two outer walls
3. the wall has to incline slightly inwards as it goes higher

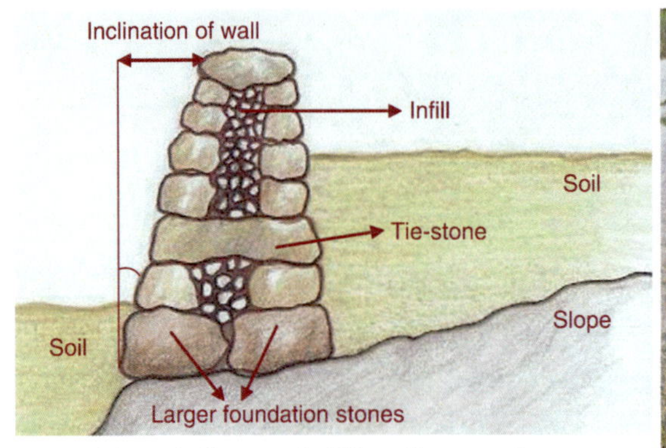

Schematic diagram of rubble wall construction to obtain terracing

Rubble wall infill, 'mazkan' (Wied Għomor, Malta)

A rubble wall is constructed as a double wall with an infill of tightly packed smaller stones, called *mazkan*, in the centre. This gives strength to the wall, preventing it from collapsing inwards. The strength and stability of the wall also depends on proper foundations which, ideally, are laid directly on bedrock. The walls are built up to the desired height layer by layer, and at intervals, large tie-stones are placed which span both faces of the wall.

The building technique is mostly focused on protecting the wall from giving way under the pressure of surface runoff during rainfall events. Apart from strengthening the wall, the central infill is also extremely important in regulating the drainage of rainwater around and inside these walls. According to Ellul (2005), the stones used for the infill must be graded and layed down systematically with the larger stones placed at the base.

Additionally, flow holes can often be seen at the base of these rubble walls. Primarily these flow holes prevent the collapse of the walls by relieving the pressure of heavy storm water. They are mostly used in walls retaining clayey soils, since these soils tend to hold more water and expand, thus increasing the pressure on the surrounding walls.

Flow holes in a rubble wall (Wied Qirda, Malta)

Detail of flow hole (Wied Qirda, Malta)

The building of longer walls, or those retaining terraces, requires more skill. Ellul (2005) notes that longer walls are usually constructed of independent "V" sections locally known as *posta* and *ġwienaħ*. Therefore if a *posta* collapses the adjacent section will not because it is not directly connected to it. Similarly, for walls retaining terraces, special considerations need to be taken with respect to the foundations, the infill and the inclination of the wall.

It can be noted that the building of these rubble walls requires expertise and skill, especially in the construction of walls retaining terraces. This traditional skill, which has been passed on over generations is unfortunately slowly dying out.

Rubble walls made up of 'Posta' and 'Ġwienaħ' (Wied il-Kbir, Malta)

Acknowledging this, in 2004, local government together with the Building Industry Consultative Council (BICC) launched a specialised training course in rubble wall building. Presently, these classes are being offered by the Malta Employment and Training Corporation (ETC).

The Many Faces of Rubble Walls

The specific characteristics of the traditional rubble walls are such that they not only support the agriculture for which they are built, but in doing so they also minimise human impact and in many aspects help to improve the local natural resources. The different beneficial aspects of rubble walls can be appreciated through the many "faces" they adopt as land developers, soil formation supporters, water collectors and walls of life.

Land Developers. The northern part of Malta and Gozo has a hilly topography. This makes agriculture hard since slopes increase water run-off and soil erosion, leaving the land depleted of its natural resources and difficult to manage. Slopes with an inclination higher than 15% are considered to be agriculturally challenging. Figures obtained from a slope map of Gozo show that 58% of the island has a slope of 15% or higher. Without the use of rubble walls and terracing this terrain would be considered as too demanding and tough to cultivate!

Table 1 Surface area of agricultural fields in Gozo in relation to slope categories

Slope Range Categories (%)	0–4	4–8	8–15	15–25	25–45	>45
Agricultural fields with rubble walls (km^2)	0.40	0.80	2.20	3.50	5.20	7.90

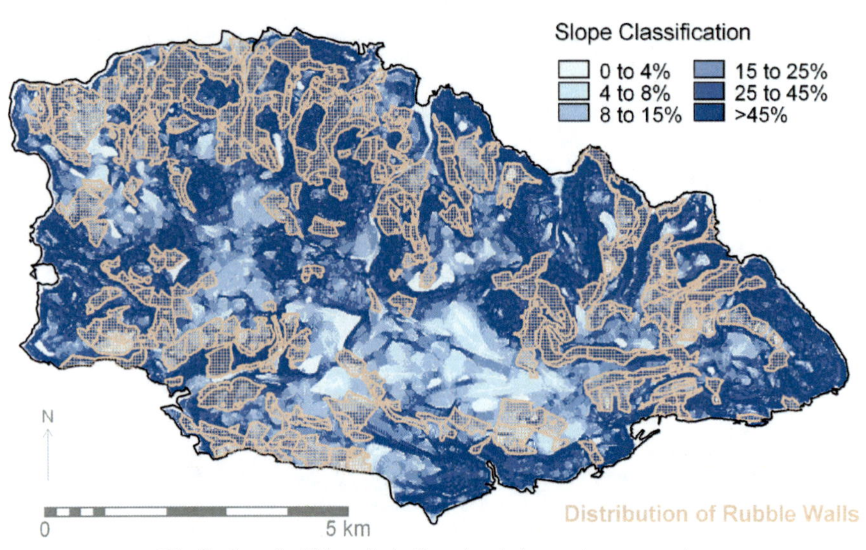

Distribution of rubble walls in Gozo in relation to slope categories

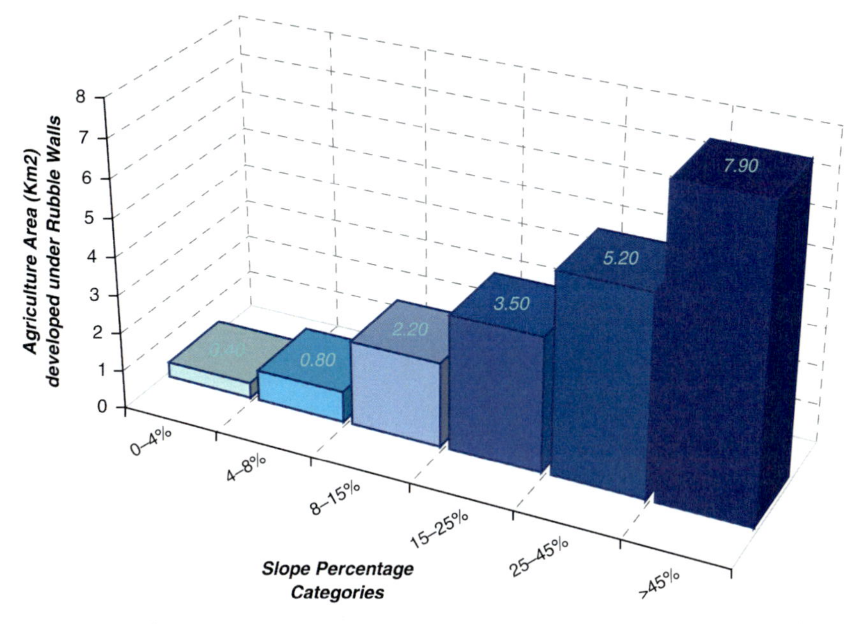

Graphical representation of figures presented in Table 1 categories

Agriculture is important for the economy of Gozo. In fact, 48.4 km^2 is arable land, which represents 72% of the island. Nearly half of this arable land, 20 km^2, is taken up by terraced fields supported by rubble walls. The majority of these rubble walls are found along the steepest slopes on the island. This highlights the importance of rubble walls in *maximising the surface area available for agriculture* in hilly areas.

Soil Formation Supporters. A longer-term function of rubble walls is to assist in the process of soil formation which in turn increases the local agricultural productivity. Soil productivity depends on specific soil characteristics such as texture, organic matter content, conductivity, pH level and soil moisture. When productivity is related to agricultural production, these are referred to as the edaphological properties of the soil.

After a rubble wall is built, the space created between the wall and the slope is refilled with the soil previously removed. Initially, the edaphological characteristics of this newly displaced soil are not yet well developed. The process of soil formation requires time for the soil horizons to reach the best edaphological characteristics. Rubble walls support the conditions required for soil formation in that they allow for crops to be cultivated and help in retaining soil moisture. In the long-term, the inputs of crop residues (roots, stubble etc.) and natural fertilizers build up the characteristics of the soil, consequently increasing its productivity. Therefore, if the field created behind the rubble wall is well managed, the edaphological characteristics and the agricultural productivity of the soil can be greatly improved over the years.

Stubble, an important agricultural input for soil formation (Wied Għomor, Malta)

Water Collectors. The role of rubble walls in relation to the hydrological cycle is also worth considering. As mentioned earlier, most of the rubble walls are found on slopes greater than 15%. In areas with steep slopes, rainwater runs quickly down the slopes allowing little time for the runoff water to infiltrate. When terracing is used, previously sloping areas are transformed into a number of horizontal areas delineated by rubble walls. Both the terracing and the rubble walls themselves slow down the surface runoff allowing the water to infiltrate into the soil. Consequently, the soil is kept humid for a longer period of time.

It follows from the above that rubble walls may also be helping in the recharge of groundwater aquifers. The only geological formation that can be considered impermeable is the Blue Clay. Thus, in terraced fields found on the Blue Clay formation, water infiltration practically stops at soil level where the water is retained in the soil as discussed above. However, when these fields are found on the highly permeable Upper Coralline, water is allowed to infiltrate deeper to recharge the underlying aquifer. In both scenarios, rubble walls are retarding the flow of surface water runoff, which would otherwise be lost into the sea. Thus by *increasing the infiltration capacity* of the terraced areas, Rubble walls are also increasing the natural renewable water resources of the area.

Walls of Life. Rubble Walls built using traditional methods and materials provide an important habitat for a variety of local flora and fauna. In rainy winters rubble walls are wet to the core. During drier periods, the infill, *mazkan*, is able to retain moisture for a long period of time, providing a varied environment with different gradients of humidity, ranging from very damp at the bottom of the wall to very dry at the top (Ellul, 2005). This allows for different species of flora to grow and establish their roots alongside and within the cavities of the rubble walls.

The plants growing on or beside rubble walls are mostly the more widespread species of herbs, shrubs and trees of the Mediterranean region. To mention a few; plant species include the Maidenhair Fern, "Tursin il-bir", (*Adiantum capillus-veneris*), the Sweet Alison, "Buttuniera" (*Lobularia maritime*), Spiny Asparagus, "Spraġ Xewwieki", (*Asparagus aphyllus*), and the Caper, "Kappara", (*Capparis orientalis*), whilst the Fig Tree, "Siġra tat-tin", (*Ficus carica*), and the Prickly Pear, "Bajtar tax-Xewk", (*Opuntia ficus-indica*) are often seen growing close to these walls.

Caper, 'Kappara' Spiny asparagus, 'Spraġ Xewwieki'

The natural fissures and cavities found in between the rough stones used to build these walls also provide shelter for many small animals. Ellul (2005) provides an interesting list of fauna that inhabit or make use of these rubble walls for various reasons. Among others, the list includes the Maltese Wall Lizard that lives in these walls, the Gecko which uses the wall to live and to build its nest, the Edible Snail which is attracted to the cool dampness of the wall in summer, the Spider which looks for food in this wall and the Weasel which nests in these walls.

Moorish Gecko, 'Wiżgħa tal-kampanja' Edible Snail, 'Għakrux raġel'

Furthermore, as the Rural Strategy Topic Paper (2003) points out, rubble walls surrounding fields provide a connection between distinct natural habitats. This means that rubble walls act as passageways for small animals allowing them to travel safely between the different habitats, away from the dangers of roads, hunters and predators. According to the Rural Strategy Topic Paper (2003) "Only few distinct areas characterised by natural habitats are connected with each other: this hinders enhancement and natural regeneration of natural habitats". Hence, rubble walls are supporting and preserving the local fauna and flora.

Silent Fighters Against Soil Erosion

The most important role of rubble walls is undoubtedly to prevent soil erosion. Soil erosion has always been a major concern and a threat to the sustainability of the local agricultural sector. In the Maltese Islands soil is a limited resource, and except for the fertile agricultural land found in valleys, soils are usually very shallow ranging in depths from 20 to about 60 cm. Additionally, a recent increase in urban development, especially on the mainland, has led to the loss of important natural and agricultural land, with the consequent loss of valuable natural habitats and precious soils.

The soil erosion process is related to soil fertility loss and vegetation cover reduction, both of which play a central part in the larger framework of the process of desertification, which in turn leads to a devastating consequence: irreversible resource loss. Furthermore, given that the Maltese Islands have a typical Mediterranean karst setting characterized by young and shallow soils, it is important to point out that soil erosion usually results in more dramatic impacts in karst environments.

Soil conservation practices such as the use of rubble walls can be employed effectively to combat the processes leading to soil erosion, and eventually desertification.

To appreciate how rubble walls combat soil erosion we look at the parameters that are involved in this process. The Universal Soil Loss Equation (Wischmeier, 1978), is the most comprehensive technique available to estimate cropland erosion at the field level. The soil erosion rate (A) is calculated using six major factors, namely rainfall erosiveness (R), soil erodibility (K), slope steepness (S), slope length (L), cropping management techniques (C) and supporting conservation practices (P): $A = R \times K \times S \times L \times C \times P$.

Terraced fields cutting through a steep slope (Xlendi Valley, Gozo)

One of the principal factors affecting soil erosion is rainfall (R), even more so if we consider the Mediterranean climate which usually brings sudden and heavy showers in autumn when vegetation cover is scarce. When rainfall reaches the ground, raindrops hit the soil surface producing minuscule cracks, the magnitude of which depends on the soil erodibility (K). The broken soil particles are now more prone to be carried away by the surface runoff produced during rain storms. The amount of soil carried away in such conditions is also affected by the slope factor (S) and the distance covered down the slope (L). The USLE equation implies that the higher the value of any of the factors, the higher the soil erosion rate. Conversely, the reduction of the value of any of the factors through cropping management techniques (C) and conservation practices (P) will result in a decrease in the rate of soil erosion. Conservation practices, such as terracing and rubble walls, do not only affect the P values in the equation, but are also decreasing the topographical factors of slope gradient and length, two very important parameters for the calculation of the soil erosion rate at the field level!

The USLE estimate is intended to highlight the importance of soil conservation practices and to enable "farmers and conservation advisers to select combinations of land use, cropping practices, and soil conservation practices, which will keep the soil loss down to an acceptable level – in today's terms it would be said to ensure that the farming system is sustainable" (Hudson, 1993). The first local farmers who built and maintained terraces and rubble walls clearly used their knowledge of the land to tackle the soil erosion problem in a very effective manner. Sadly, today, very few people recognize the importance of these conservation practices, and most of the terraced fields and rubble walls are not adequately maintained, putting at risk a precious resource which has been painstakingly protected for several hundreds of years.

Soil slumps are observed in the upper terrace. Lower terraces show better maintained rubble walls (Wied Għemieri, Malta)

A field survey of rubble walls in Gozo carried out by IRMCo in 1999 in the context of the EU-sponsored International Cooperation (INCO) Project entitled *ResManMed:* Resource Management in Karstic Areas of the Coastal Regions of the Mediterranean, proved very enlightening in assessing the relationship between the condition of the rubble walls and the areas affected by soil erosion. Rubble walls were classified according to their condition, i.e. good, medium or bad. In parallel to this, the occurrence of soil slumps at the field level was recorded according to the following categories: isolated or abundant.

The occurrence of soil slumps correlates extremely well with the condition of the rubble walls on the island. The map clearly reveals that rubble walls in poor condition were found predominantly in areas where a high incidence of soil slumps had been recorded. Walls in good condition are associated with practically no occurrence of soil slump. Both illustrations highlight the significance of well-maintained rubble walls as a conservation practice to combat soil movement and soil loss.

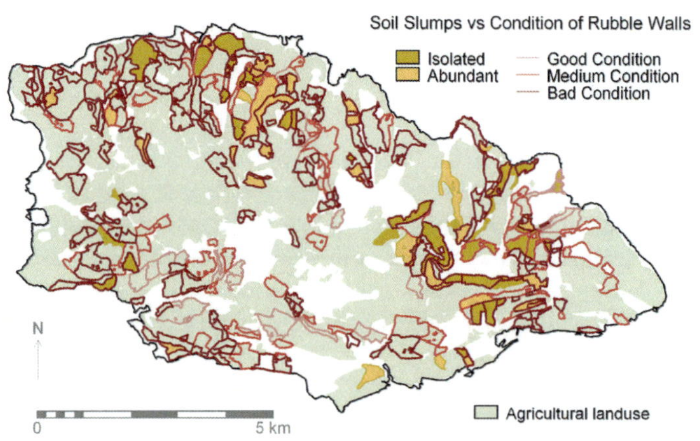

Occurrence of soil slumps in relation to the condition of rubble walls in Gozo

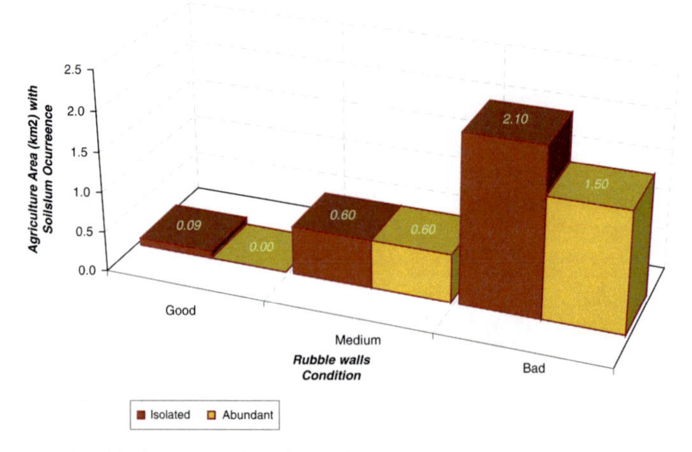

Graphical representation of the information presented in the above Map

In recent decades, there has been a steady loss of rubble walls, mostly because of the loss of land due to development (buildings, roads, quarries etc.). At the same time, traditional knowledge is being lost with the decline in the local farming community. While legislation does require that these walls are adequately maintained, it is unfortunate that a poor condition of these walls can be witnessed in many areas in Malta and Gozo. The Rural Strategy Topic Paper (MEPA, 2003) acknowledges that bad practices in wall maintenance have resulted in the loss or damage to traditional rubble walls. These bad practices include the addition of new materials or new building methods without the proper consideration or understanding of the important functions of the rubble walls. The document insists on the importance of providing further control and guidance with respect to the maintenance, rebuilding and construction of traditional rubble walls.

The rubble walls have been providing us with huge benefits throughout the centuries

Let's take good care of our *ancestral country allies!*

References

Borg, J.J. et al. 2007. 'Nature in Gozo'. Birdlife, Malta.
Ellul, E. 2005. Il-Ħitan tas-Sejjieħ. Klabb Kotba Maltin, Malta.
Government of Malta 1973. Fertile Soil (Preservation) Act (1973) Chapter 236 found at http://docs.justice.gov.mt/lom/legislation/english/leg/vol_5/chapt236.pdf accessed 1/8/2008.
Government of Malta 1997. Rubble Walls and Rural Structures (Conservation and Maintenance) Regulations – LN 160, 1997, (amended LN 169, 2004). Malta.
Haslam, S. M. & Borg, J. 2002. Let's Go and Look After Our Nature, Our Heritage!. Ministry of Agriculture & Fisheries – Socjeta Agraria, Malta.
Hudson, N. W. 1993. Field Measurement of Soil Erosion and Runoff. Silsoe Associates, UK. FAO, Rome found at www.fao.org/docrep/t0848e/t0848e-12.html#TopOfPage accessed on 6/8/2008.
MEPA 2002. National report on the implementation of the United Nations Convention to combat desertification. MEPA, Malta.
MEPA 2003. Rural Strategy Topic Paper – Volume 1, Final Draft found at http://www.mepa.org.mt/planning/factbk/SubStudies/RuralTP/Rural_TP.pdf accessed on 1/8/2008
MEPA 2006. State of the Environment Report for Malta – Sub Report 4: Land'. MEPA, Malta.

Ministry for the Development of Infrastructure, Planning Division, 1990. Structure Plan for the Maltese Islands. Malta.

ResManMed, unpubl. Interim and Final Reports on the INCO-DC project ERBIC18CT970151, Resource Management in Karstic Areas of Coastal Regions of the Mediterranean, 1997–2000.

Sammut, S. 2005. State of the Environment Report for Malta – Background Report on Soil. MEPA, Malta.

Wischmeier, W. H. & Smith, D. D. 1978. Predicting Rainfall Erosion Losses – a guide to conservation planning. U. Department of Agriculture, Agriculture Handbook No. 537, U.S.

Cappadocia (Kapadokya)

Ferika Özer Sari and Malike Özsoy

Cappadocia is located within central Anatolia between the cities of Nevşehir, Kayseri and Niğde. Three volcanoes on the Nevşehir plateau, Mount Erciyes (3,917 m), Mount Hasan and a smaller mountain named Güllüdağ had formed this territory by frequent and continual eruptions and earthquakes for thousands of years in the Tertiary period.

Situation of Cappadocia on the Map of Turkey (Turkish News Agency, 2003)

Ten million years ago, between the late Pliocene and early Pleistocene, these three volcanic mountains left the region thickly layered with a soft, porous stone known as tuff, formed from hot volcanic ash (Cookson, 2007). The Cappadocian ignimbrite (fine-grained volcanic rock mainly composed of welded fragments of quartz and feldspar) succession of central-southern Anatolia comprises at least nine major and two minor calc-alkaline rhyolitic sheets (Piper et al., 2002). This

F.Ö. Sari (✉)
Yasar University, 35500 Bornova-İZMİR
e-mail: ferika.ozersari@yasar.edu.tr

N. Evelpidou et al. (eds.), *Natural Heritage from East to West*,
DOI 10.1007/978-3-642-01577-9_32,

Andesitic layer of tuff was in turn overlain by a series of andesitic and basaltic lavas. The lava and ashes piled up and created blocks of soft volcanic rock hills having relatively thinner, hard basalt layers in some parts.

Since then, wind, rain and the Kızılırmak river have eroded the rock to create a spectacular lunar scenery (Cooksan, 2007). The deep layers of soft, quite homogenous material have been eroded to form regular conical peaks or irregular masses reflecting their differential resistance to weathering. Typical geomorphological features include pillars, columns, towers, obelisks and needles, sometimes reaching a height of 40 m.

These pillars with chimney caps, have been named "Fairy Chimneys" (Umar, 1998). Eroded volcanic deposit layers also formed a series of interconnected valleys and ravines, dramatically punctuated by plateaus and strange, conical outcroppings (Lévy, 1997).

Cappadocia has been continuously inhabited from the pre-Christian era up to present day. It was known as Hatti in the late Bronze Age, and was the homeland of the Hittite power centered at Hattusa. After the fall of The Hittite Empire, Cappadocia was left in the power of the Persian Empire. Monasteries, hermitages, shrines and even dwellings were carved in the tuffs. The carved grottoes reflect Byzantine architecture, style and painting. Most of the dwellings and chapels date from the tenth to thirteenth centuries.

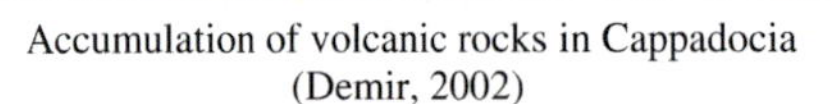

Accumulation of volcanic rocks in Cappadocia (Demir, 2002)

Fairy Chimneys in Pasabaglari, Zelve (Demir, 2002)

Paşabağları Chimneys (Demir, 2008: 60)

Fairy Chimneys from Ürgüp

During the Middle Ages, when Cappadocia was an important province of The Byzantine Empire, it became a vibrant area of habitation. The lack of green vegetation (and wood) and frequent earthquakes forced the inhabitants to make strong, safe dwellings. The soft rock of Cappadocia was excavated deep under the ground and at the same time, carved above ground and used as shelter by human beings in ancient times (Turkish Ministry of Culture and Tourism, 1990).

Uchisar Citadel (Demir, 2008: 47)

Inner view from Saint Barbara Church (Demir, 2002)

Hundreds of churches and monasteries are carved into the rocky landscape. More than seven hundred churches alone have been counted in the region many of them preserving impressive ensembles of fresco decoration (Lévy, 1989).

The Valley of Göreme is a unique area where human activity has blended unobtrusively into the landscape (Hudman and Jackson, 2003). This valley of churches, and monasteries is located about 15 km from Nevşehir and called the Göreme Open Air Museum.

A group of churches carved into rock at Göreme (Umar, 1998)

Saint Barbara (Ayia Varvara) church is one of the hundreds of churches carved into the rocks. This ninth century church of the iconoclastic years has two pillars

Christ Pantocrator, Karanlık (Dark) Church, (Demir, 2008: 29)

Stairs to lower decks of "Derinkuyu Underground City" (Demir, 2002)

and a mini dome. On the walls, geometric figures of reddish earth colors can be seen. The paintings of Christ, St. George, St. Theodore and St. Barbara are from a later period (Umar, 1998).

Subterranean sites are another aspect of cultural and natural heritage. Underground cities and semi-underground settlements, most of which are at least 1,500 years old, exist in the Cappadocia Region of Turkey (Aydan, 2003). Explanations about "when, why and by whom" the subterranean sites were created are diverse. According to some specialists, the underground cities were created, and used as storage areas by the Hittites and were much later enlarged and used as refuges by Christians who were tormented by the Romans. Some others propose that these cities were created at a later time, by the Phrygians, as shields or a line of defense against the Assyrians (Demir, 2002).

No one knows how many underground cities lie beneath Cappadocia. Eight have been discovered, and many smaller villages, but there are doubtless more. The biggest, Derinkuyu, was discovered as recently as 1965, when a resident cleaning the back wall of his cave house broke through a wall and discovered behind it a room that he'd never seen, which led to still another, and another.

Eventually, cave archeologists found a maze of connecting chambers that descended at least 18 stories and 280 feet beneath the surface, ample enough to hold 30.000 people – and much remains to be excavated. One tunnel, wide enough for three people walking abreast, leads to another underground town six miles away. The existence of other passages suggests that at one time all of Cappadocia, above and below the ground, was linked by a hidden network. Many still use the tunnels of this ancient subway system as cellar storerooms (Ousterhout, 2005).

Bringing together the best of the Tertiary and the Byzantine periods, the combination of scenic geological wonder and arcane art history has made Cappadocia a tourist destination of ever increasing popularity (Ousterhout, 2005).

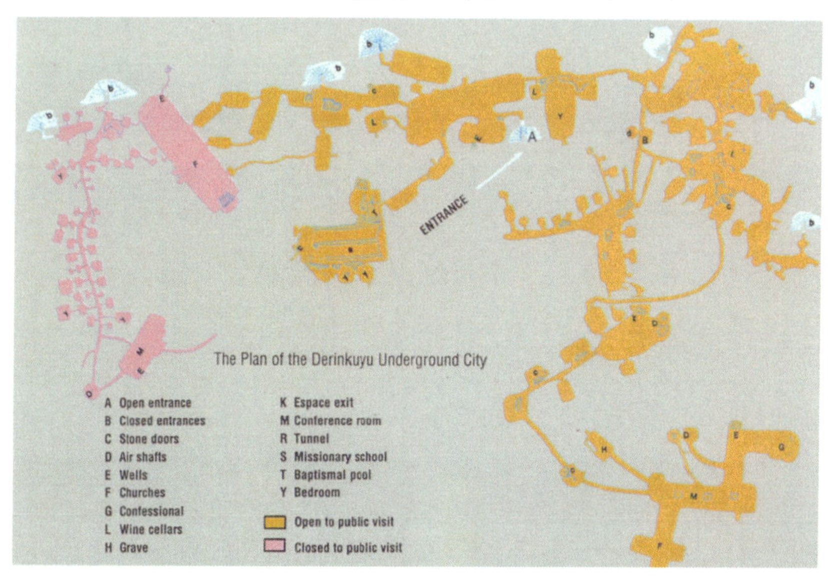

The Plan of Derinkuyu Underground City (Demir, 2008: 68)

References

Aydan Ö., Ulusay R., (June 2003). *'Geotechnical and geo environmental characteristics of man-made underground structures in Cappadocia, Turkey'* Engineering Geology Volume 69, Issues 3–4, Pages 245–272.

Cookson L., (28 July 2007). 'Away with the fairies: Cappadocia is full of magical natural attractions' Newspaper.

Demir Ö., (2002). Kapadokya, Medeniyetin Beşiği, 9th Edition, Promat A. Ş., İstanbul.

Demir, Ö., (2008). Cappadocia, Cradle of History, 12th Revised Edition, Pelin Ofset, Ankara.

Geoff, M., (August 2007). 'Derinkuyu or the Allure of the Underground City' BLDGBLOG Architectural Conjoncture Magazin.

Hudman L., Jackson R., (2003). Geography of Travel & Tourism, 4th Edition, Thomson Delmar Learning, NY.

Levy J., (1989). Eglises byzantines; KOSTOF S, 'Caves of God: Cappadocia and Its Churches', 2nd Edition, Clarendon Press, Oxford.

Levy J., (1997). 'Cappadoce: Mémoire de Byzance'. CNRS éditions, Paris.

Ousterhout R., (2005). 'Byzantine Settlement in Cappadocia' Washington USA.

Piper J D A., Gürsoy H., Tatar O., (October 2002). 'Palaeomagnetism and magnetic properties of the Cappadocian ignimbrite succession, central Turkey and Neogene tectonics of the Anatolian collage' Journal of Volcanology and Geothermal Research, Volume 117, Issues 3–4 Pages 237–262.

TURKISH MINISTRY of CULTURE and TOURISM, (1990). Generale Directorate of Information, Göreme Nature's Mysticism.

TURKISH NEWS AGENCY, (2003). Turkey 2003, Directorate General of Press & Information of the Prime Ministry, Ankara.

Umar B., (1998). Kappadokia, Bir Tarihsel Coğrafya Araştırması ve Gezi Rehberi, Tükelmat A.Ş., İzmir.

Caves of Turkey

Muhammed Aydoğan

Currently it is believed that civilization started with agricultural settlements and that immigrating groups of people had a primitive way of life within their natural environment. According to this point of view, caves have been accepted as habitats which reveal good traces of prehistoric life. However, taking a closer look at the beginning of the process, it can be clearly seen that caves were not only used by primitive people as functional (and advantageous) shelters, but have also been used by the first agricultural societies and even by urban societies following them. Thus caves can be seen as shelters which reveal detailed information about the development of various cultures.

Entry of a natural (karst) cave, near the the Taskale – Karaman Road

M. Aydoğan (✉)
Department & Department of GIS, Faculty of Architectural, Dokuz Eylul University, City and Regional Planning
e-mail: m.aydogan@deu.edu.tr

N. Evelpidou et al. (eds.), *Natural Heritage from East to West*,
DOI 10.1007/978-3-642-01577-9_33,

Caves are also a good source of very valuable information about natural formations – especially geological ones. Such formations are studied not only in speleology, but also by various sciences such as geomorphology, geology, hydrology-hydrogeology, meteorology, biology, archaeology, and anthropology. Besides their scientific significance, caves are also important for tourism, the local economy and city planning.

Although massive amount of research has been carried out by many groups at various universities and institutions, it is obvious that there are still many unexplored caves in Turkey, (mainly because of lack of accessibility). Despite all this documentation prepared by different groups, there is as yet no comprehensive study or inventory. The caves of Turkey present distinctive samples of different types and they reveal traces of different cultures of different ages in their rich stratified deposits. The purpose of this study is to introduce various types of caves in Turkey by classifying them according to their interaction with different cultures.

Types of Caves

Caves are described as underground voids large enough for a human to enter. These can be a couple of meters or kilometers long, and from a few meters to hundreds of meters deep (MTA, 2008). Some suggest that the term cave should only apply to cavities having some part that does not receive daylight. However, in popular usage, the term includes smaller spaces like sea caves, rock shelters, and grottos, (Wikipedia, 2008).

The generally accepted definition of a cave is a cavity that is a natural formation of various geological processes. But some caves contain evidence of human

Zonguldak, Gokgol Cave

habitation dating back hundreds of thousands of years. Human interference can be seen in "karst" areas, and especially in caves of limestone and chalk. In caves where human habitation is common, only experts can differentiate which caves were naturally formed and which were reshaped by human effort. In Anatolia, for this reason, the famous underground cities, even the underground depots (used like climate controlled storage areas) dug out by people are called caves. In this paper, all types of caves are taken into consideration. Those made by humans are classified as "artificial" caves. This definition is in accordance with the definition of a cave (MTA, Mine Technical Research Institution, 2008).

Natural caves usually have the form of series of cavities forming a network. Some of these cavities are large enough for a human being to enter, some of them contain very narrow cracks and channels, and some others enormous spaces.

Rock sample from Burdur Insuyu Cave

Caves can be classified in two sub-groups according to their phase of formation. Those formed during the formation of the main rock (like lava caves, glacial caves and holes in travertine) are called "primary caves". Caves formed after the formation of the main rock are called "secondary caves". In caves of this sort, it is underground waters containing carbonate, sulfur, or chlorine that generally erode rocks (MTA, 2008).

Different kind of geological processes create various cave forms, a mechanism known as speleogenesis. This may involve a combination of chemical processes, erosion caused by water, tectonic forces, microorganisms, pressure, atmospheric influences, and even digging. Most caves are formed in limestone by dissolution (Wikipedia, 2008). Atmospheric water, rich in carbon dioxide (CO_2), turns to carbonic acid (H_2CO_3) while it passes through the cracks of carbonated rocks. This extremely solvent and acidic water mixture erodes its course and forms "karst" caves (MTA, 2008).

Burdur Insuyu Cave, Rock Samples

Many of the natural caves are closely related to underground water because of chemical effects and dissolvent attribute of water. In most caves of this type, pools of different sizes can be found. For example, there are 15 pools in Burdur Insuyu cave and people can even swim in some of them. In a very few caves there may be underground running water.

Running underground water in Zonguldak Gokgol Cave

Underground water running from a cave (Kayseri Kapuzbasi)

In natural caves untouched by humans, there may be stalactites, stalagmites and columns of different lengths, forms and colours. Stalactites are formed by the deposition of calcium carbonate and other minerals, precipitated from mineralized water solutions. A stalagmite is a type of speleothem that rises from the floor of a limestone cave due to the dripping of mineralized solutions and the deposition of calcium carbonate. The combination of stalactites and stalagmites into a single form is called a column.

Stalactites, stalagmites and columns in Zonguldak Gokgol Cave

Stalactites, stalagmites and columns in Gumushane Karaca Cave

Natural Caves in Turkey

"Majority of the caverns in Turkey is a part of ground/underground formations which is called "karst" in geomorphology. Caverns represent only a specific scale group of this wide spectrum of karst formation typology" (TAY (Turkish Archaeological Settlements), 2008).

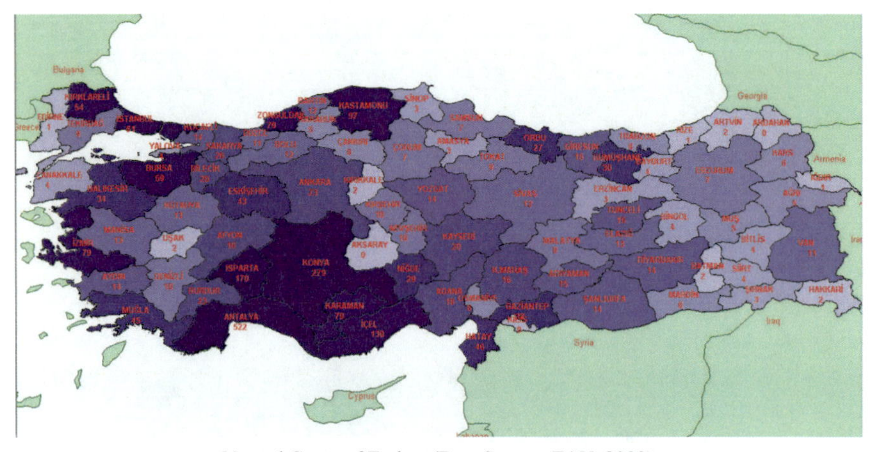

Natural Caves of Turkey (Data Source: TAY, 2008)

Unfortunately there is no inventory giving the actual number of caves/caverns in Turkey. According to TAY's web-based GIS database, which is designed to help coordinate academics working on archaeological sites, there are 2,443 caves/caverns in Turkey, (TAY, 2008). The number of identified caves by MTA is around 800, (MTA, 2008). The Ministry of Culture and Tourism claims that in Turkey, where 40% of rock formation is Karst, there are more than 20,000 caves. Again, the statistics of the Ministry show only 1,250 caves (6.25%) have as yet been discovered, (including those 800 caves identified by MTA).

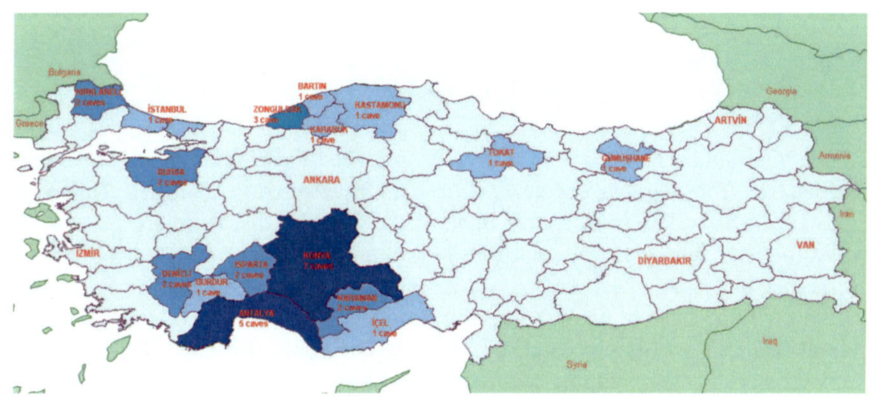

Caves of Turkey prepared for tourism by the Ministry of C&T.
(Data Source: Ministry of Culture & Tourism, 2008)

The longest cave in Turkey is Pinargozu Cave in the province of Isparta with a length of 16 km, and the deepest is the Peynirkonu Cave in Mersin with a depth of 1,453 m, (Ministry of Culture and Tourism, 2008). Only thirty three of these natural caves/caverns are open to tourism and managed by the Ministry. Some of these caves are open to public, whereas some are open only to visitors with adequate equipment or for professional visits or research. (Table 1)

Table 1 Caves of Turkey, prepared for tourism by Ministry of Culture & Tourism

Open to public visitors			Open only to visitors with adequate equipment			Open only to professional visitors with adequate equipment		
District name	Cave name	Lenght (Metre)	District name	Cave name	Lenght (Metre)	District name	Cave name	Lenght (Metre)
Antalya	Damlataş	200	Antalya	Yerköprü	500	Antalya	Altınbeşik	2200
Antalya	Dim	375	Bursa	Oylat	665	Isparta	Pınarözü	16000
Antalya	Zeytintaşı	136	Bursa	Ayvaini	5500	Karaman	Gürleyik	2500
Bartın	Gürcüoluk	169	İçel	Gilindere	555	Konya	Sakaltıtan	303
Burdur	İnsuyu	525	İstanbul	İkigöz	4816	Konya	Susuz Güvercinlik	1351
Denizli	Kaklık	190	Karaman	İncesu	1356	Konya	Pınarbaşı	800
Denizli	Dodurgalar	145	Kastamonu	Ilgazini	860	Zonguldak	Kızılelma	6630
Gümüşhane	Karaca	256	Konya	Baltaini	1768			
Isparta	Zindan	760	Konya	Körükini	1250			
Karabük	Mencilis	6052	Konya	Suluin	300			
Kırklareli	Dupnisa	2750	Kırklareli	Yanasu	1620			
Konya	Tınazteper	1650	Zonguldak	Çayıköy	1004			
Tokat	Ballıca	1180						
Zonguldak	Gökgöl	3350						

Data Source:
(Kültür ve Turizm Bakanlığı, 2008)

Other cave-like formations are "sinkholes", which can be found in many locations in Turkey. A sinkhole, also known as a sink, shake hole, swallow hole, swallet, doline or cenote, is a natural depression or hole in the surface topography caused by the removal of soil or bedrock, often both, by water (Wikipedia, 2008). Sinkholes can be formed in ponds retaining large amounts of rain. Some sinkholes are dry but many look like small lakes or ponds (Kocbay and Kilic, 2006). These sink holes are generally used for agricultural irrigation rather than for tourism.

Different Use of Caves

Use of Caves for Early Human Habitation

It is quite natural that early primitive people used caves and caverns to protect themselves from the dangers of wild life and atmospheric conditions. They were incapable of building even simple habitats. It is known that there are many caves used for early human habitation in Turkey, although the best-known examples are in France and Spain. Many of the traces of early human settlement in caves have disappeared because of natural abrasion and use in later periods. Despite this fact, completed and ongoing research on many caves proves that caves have played a significant role in early human settlements in Anatolia.

Karaman Manazan Caves

The majority of caves in karst formation in Anatolia were used as habitats from the sub-Paleolithic age (about 160,000 years ago). The most famous examples are Yarimburgaz, Karain, Ucagizli and Okuzini Caves (Kirman and Ulusoy, 2005). Radio-carbon analysis dates Karain Cave back to 39,630 years (Kartal, 2006).

Fireplace and grain stores in Karaman Manazan Caves

It is understood that there was cultural differentiation in prehistoric society, and that various social groups used caves for different purposes. For this reason, caves are places where much anthropological evidence can be found. For example, Yarimburgaz Cave in Istanbul provides evidence of living conditions dating back 7,000 years (Meric et al., 1991).

Usage of Artificial Caves as Hidden Habitats

With the experience of utilizing natural caves as habitats, man started to form artificial caves by digging or carving rocks, or enlarging existing small caves where there was no suitable natural caves/caverns. The use of these caves has changed over time and people have started to use them as grain warehouses, animal barns, hideouts or fresh vegetable/fruit freezers.

Sumela Monastery in Trabzon Province

For different functions, people enlarged the volume of caves, added more rooms, separated them into sections and created entrances by carving the rock surface.

Traces of fireplaces, grain storage, and utilities such as wells provide evidence that these caves were used for habitation rather than simply for storage. These kinds of caves are generally connected to each others by horizontal corridors or vertical tunnels, and have doors and sometimes windows. These features are evidence of a society living in caves.

Another interesting feature of these caves is their rather small entrance, which is often very difficult to locate from a distance, despite their massive spaciousness and number of sections under the ground. These entrances tend to be located at height, covered with plants, trees or bushes. The difficulty of finding entries from a distance together with their location on high cliff edges that hard to reach, indicates that people had serious fears of various dangers in the past.

This situation is displayed in a different way in Nevsehir, another settlement in Central Anatolia, or Cappadocia. There are underground cities in this region dug in tuff surface. These underground cities are formed of many continuous and linking rooms and the form can be recognised as a different type of a cave system. There are many unexplored underground cities in this region, besides the ones explored such as Kaymakli, Derinkuyu, Mazi, Ozluce, Ozkonak, Sivasa Gokcepinar, Tatlarin and Acipinar. The earliest use of these cities is accepted to be in the first century after Christ. The early Christians of Anatolia used these cities to protect themselves and their religion from Roman despotism (Municipality of Kaymakli, 2008). It is estimated that they were also used before the Christian era, during periods of clashes resulting especially from power struggles between different cultures. But because of interference by the dense Christian population, and the usage of these places for storage in the following centuries, the traces of earlier centuries were largely lost.

Usage of Natural Caves as Hiding Places

Because of the Roman oppression on early Christians, these people used various methods of hiding in different regions of Anatolia. In regions where many natural underground caves existed, such as in Cappadocia, it can be seen that they used these natural caves and furthermore they built walls within them. The caves in Aksaray Ihlara Valley and Cennet (Heaven) and Cehennem (Hell) Caves in Mersin (the sunken caves) are the best well known examples.

When Christianity became widespread in Anatolia, some of these caves were used for concealment. Religious locations were accepted as holy places and churches, or monasteries, were built within them. The Sumela Monastery in Trabzon is regarded as one of the most distinctive and genuine of all these structures; it was built like a castle in the entrance of a large natural cave in a cliff.

Interior of the Sumela Monastery

Usage of Artificial Caves for Agrarian Storage

Although variations in humidity depend on the characteristics of each cave, the temperature inside natural caves is stable during all seasons. Thus caves have always been used for the storage of various agrarian products. However, because of the difficulties of transportation of agricultural products to natural caves, and for greater security, people started digging caves, mostly in rural parts of Central Anatolia. These artificial caves are commonly dug between farm buildings and houses; they belong to families and the stored product can be shut up behind a door.

Artificial caves for agrarian storage, in Karaman Taskale

The caves in (Karaman) Taskale are accepted as one of the most outstanding examples of agrarian storage caves. They are on the top of one another on a steep cliff, and people climb up by holding onto the carved rock to reach to the top. There are wooden cranes with pulleys over doors of the storage areas which are used to pull the product up to the caves. Although it is well known that these caves have been used for the same purpose for centuries, their date of first usage is not clear because of the abrasion of the rock.

Antique derricks and the doors of artificial caves for agrarian storage, in Karaman Taskale

As well as caves in this region that are now under protection, there are many caves being dug out for agrarian storage in many other places of Central Anatolia. They are mostly preferred for the storage of potatoes, apples and citrus fruits because of the humidity inside the caves. Most of these caves are large enough for a truck to enter and are hundreds of meters long.

Artificial caves created in last decade, for agrarian storage, in Aksaray Kavak

The Role of Caves in Culture

Cave paintings are accepted as the first cultural products of early human beings. In those times, the role of caves in cultural life was more important than their function as shelters. It is generally believed that Homer, the ancient Greek epic poet of Anatolia, wrote his poems the Iliad and the Odyssey in the caves near the ancient city of Smyrna (today's Izmir), (Kirk, 1965).

It is a fact that most caves, from earliest times, have been accepted as holy places and used for various religious rituals. Islam has been prevalent in Anatolia during the last millennium and, even in this religion, caves have a special place. Hira, the cave where the Prophet Muhammad received his first revelations from God, is still visited by respectful pilgrims. There are many other caves in Anatolia that are visited for religious purposes. The most popular ones are "Ashabi Kehf" (Seven Sleepers) caves in Izmir and Tarsus (there are 33 other caves called "Seven Sleepers" in Turkey), which are accepted as holy and are visited by Muslims and Christians alike (Akgunduz et al., 2008).

Conclusion

The objective of this study is to introduce different types of caves of Anatolia by classifying them according to their interaction with different cultures, and to be a basis for future study and creation of an inventory. The thematic map that was drawn with CBS instruments to show the locations of caves in Turkey demonstrates that tourist caves are most common in Central Anatolia and the Mediterranean region. The research may also be used for identifying the cities in which caves should be advertised and opened to tourism. But beyond all spatial analyses, owing to the fact that only 6.25% of all caves in Turkey have yet been explored and only 2.6% are open to visitors, research clearly indicates that more comprehensive studies on caves should be done in the future.

References

Akgunduz, A., Bas, Y., Tekin, R. & Kasikci, O. (1993). Arsiv Belgeleri Isiginda Tarsus Tarihi ve Eshabi Keyf "Yedi Uyurlar". In Homepage of Municipality of Tarsus. Available at: http://www.eshabikehf.org (accessed 24 July 2008).

Kartal, M. (2006). Excavations at Karain Cave in 2005. Available at: http://www.akmedanmed.com/pdf/2006_5.pdf (accessed 24 July 2008).

Kirk, G.S. (1965). Homer and the Epic: A Shortened Version of the Songs of Homer: 190. London: Cambridge University Press. Available at: http://en.wikipedia.org/wiki/Cave (accessed 24 July 2008).

Kirman, E. & Ulusoy, E. (2005). Paleolitik Donemde Dogal Yerlesim Yeri Olarak Kullanilan Anadolu Magaralari". In Proceedings of Turkiye Kuvaterner Sempozyumu TURQUA-V. Istanbul: ITU Avrasya Yer Bilimleri Enstitusu.

Kocbay, A. & Kilic, R. (2006). Engineering Geological Assessment of the Obruk Dam Site – Corum – Turkey. In Engineering Geology 87(3–4): 141–148.

Meric, E., Sakinc, M., Ozdogan, M. & Ackurt, F. (1991). Mollusc Shells Found at the Yarimburgaz Cave. In Journal of Islamic Academy of Sciences 4(1): 6–9.

Ministry of Culture & Tourism. (2008). Cave Tourism. Available at: http://www.kultur.gov.tr/TR/BelgeGoster.aspx?F6E10F8892433CFF03077CA1048A18343DB31F6A1D609AA7 (accessed 24 July 2008).

M.T.A. (General Directorate of Mineral Research & Exploration) (2008). Magaralar ve MTA. Available at: http://www.mta.gov.tr/magara/tasari.html (accessed 24 July 2008).

Municipality of Kaymakli. (2008). Yeralti Sehirleri. Available at: http://kaymakli.com/index.php?option=com_content&task=view&id=15&Itemid=32 (accessed 24 July 2008).

T.A.Y. (Archeological Settlements of Turkey) (2008). Turkiye Arkeolojik Yerlesimleri Projesi. Available at: http://tayproject.org/veritab.html (accessed 24 July 2008).

Wikipedia (2008). Cave. Available at: http://en.wikipedia.org/wiki/Cave (accessed 24 July 2008).

Dalyan Paradise

Sabah Balta

Dalyan is one of Turkey's better kept secrets, a world in itself. There are not many accessible places left on Earth where the elements have retained their primeval freshness. Contemplate Dalyan from above when you are up in the ancient town of Caunos. It is a rare piece of natural beauty.

Dalyan

S. Balta (✉)
Department of Tourism and Hotel Management, Yasar University, Izmir, Turkey
e-mail: sabah.balta@yasar.edu.tr

N. Evelpidou et al. (eds.), *Natural Heritage from East to West*,
DOI 10.1007/978-3-642-01577-9_34, © Springer-Verlag Berlin Heidelberg 2010

There is a charming summer resort area on the banks of the natural fjord like delta connecting Köyceğiz Lake to the sea; it was called Calbis in ancient times. Today its name is Dalyan. Dalyan lies within a national conservation area with unspoilt natural beauty. It is a unique place, situated on the river channel that links the vast and serene Koycegiz Lake with the stunning Iztuzu beach and is surrounded by pine-clad mountains.

Dalyan and its silent, spellbinding atmosphere are unique in the Mediterranean and Agean. Here one feels caught up in a strange and beautiful world. Spectacular scenery, the existence of biological richness such as sea, dunes, lagoons, swamps, lakes, forests, wildlife and the historic site of Caunos attract numerous visitors to this charming and beautiful town. Great clumps of reeds in the Dalyan River make a natural channel up and down which motorboats ply. For ecological balance, speed boats and boats which are longer than 12 m are forbidden to enter the river. Maximum speed for the boats in the river is 5 miles per hour.

Dalyan map

Sometimes smaller boats seem to lose their way among the reeds, because they are so dense. Dalyan is a small, charming village where life goes on much as it has for a 1000 years. One of the most beautiful sights around Dalyan are the extensive orange and lemon groves, and the fig and pomegranate trees. In addition there are spectacular channels that weave like a labyrinth, surrounded by reed beds and endless golden sands. Dalyan has excellent natural surroundings, weather, water and people, along with the delicious fish and seafood of the Mediterranean. Dalyan is one of those very rare regions in the world that still retain their natural charm and beauty.

Dalyan atmosphere

River motorboats

Dalyan is located in Muğla province and borders on the Ortaca region. The region has been designated as "Köyceğiz-Dalyan Special Protection Area" by the Council of Ministries in 1988. The areas is composed of Lake Köyceğiz with a surface area of 55 km^2, a 10 km long channel connecting the lake and the Mediterranean Sea and a delta formed where the channel approaches the sea where Iztuzu Beach takes its place in front. This is one of the few surviving places of paradise, an area of natural beauty and historical interest.

Dalyan river

The population of Dalyan is about 5300 and this may reach several times its original level during a tourism season. Dalyan has a Mediterranean climate, thus it presents the typical flora characteristics of this climate type. The region has very rich plant cover. Many observers from different parts of the world come to Dalyan to see rare birds.

Dalyan speed boats

Ancient Caunos

According to Heredotus, the father of history, even though the people of Caunos were the citizens of Caria, they considered themselves as Cretans. Strabo, the geographer, writes about the existence of a dockyard and a harbor in Caunos, whose entrance could be closed when needed.

The Aegean and Mediterranean coast of Turkey is acknowledged to be one of the best cruising areas in Europe. There are many unspoiled areas rich in natural beauty and in coastal settlements and villages it is still possible to observe local people going about their lives in traditional Turkish village fashion. Archaelogical ruins and historic sites abound in Turkey's southern coastal regions. Many ruins and sites such as Caunos, Cleopatra's Island, Knidos, Patara and Kekova can be seen directly from the sea. It is about a kilometer from Dalyan to the ruins of Caunos. Once across the river, a walk of about 800 m will take you there. The tombs hewn out of the rock in the mountainside, which are the most striking feature of Caunos, date from the fourth century BC.

Kaunos

Caunos, which was an independent state, had a unique culture. Since its foundation around 1000 BC, Caunos became famous for the trading of salt, salted fish, figs and slaves. The city sprawls over a broad slope overlooking the sea and delta. At Caunos there is an Acropolis surrounded by the city walls, a theatre, four temples, an agora, shops, harbour, Roman bath and cistern. Some of the ruins are still underground and yet undiscovered.

Tombs

Below Caunos there are clusters of rock tombs. The left hand group may be reached by walking, but to visit the right hand group you will need a rope or a ladder. One of these tombs is enriched by the relief of two lions facing each other, while the tomb farthest to the right has been left half finished, which affords us some idea of how these tombs were made. The tombs in the mountainside along the riverbank are spectacular, but one should not overlook those further away. These are Carian tombs. Now even the bones are gone but the tombs, remaining as evidence of civilizations gone by, make us think in order to understand them. Leaving Dalyan behind, we may proceed to Caunos and a time that antedates our own by 3000 years.

Caunos is thought to have been founded in the third millenium by Caunos, the son of Miletos. There are two legends that have come down to us concerning Caunos and his sister Byblis. According to one the sister was madly in love with her brother, who spurned her, whereupon Byblis hanged herself. From this sprang the expression, "Caunian love."

Caunos tombs

Because of their differing customs, Herodotus distinguishes between the Caunians on the one hand and the Carians and Lycians on the other. He also tells us that in time the Caunian language came to resemble that of the Carians, and Georgians. The second legend has it that Caunos was exiled because of his incestuous relationship with Byblis, and came to this region of Carians where he founded a city in his own name. The inscription shows that although there is a great similarity

Dalyan-iztuzu

between the two languages, Caunian has disitinctive features of its own. In order to see all of ancient Caunos in a single view you must climb to the citadel high on a hill. From here you will have a prospect of the theater, church, baths, temple, pool, Carian tombs, city walls, ancient harbor, Dalyan and Iztuzu. As the South face of the hill is a steep precipice, one must climb up via the theater side.

Iztuzu Beach

Iztuzu beach has become world famous for the gigantic turtle, Caretta caretta Iztuzu is a glorious beach on the Mediterranean, the gateway to Caunos and Dalyan. The river, anciently known as the Calbis, flows into the sea via a narrow canal some thirty meters across, which was formerly chained off by the people of Caunos to bar entrance to foreign traffic. In the time of the Caunians Iztuzu was a harbor. In ancient times, the solid part of the Dalaman River delta formation, between Bozburun and Gökbel Mountain and the point where Dalyan stream reached the sea, turned into a coast with a length of 6 km.

Caretta caretta eggs

According to geographers, the coast is a solid soil mass, developed in a lagoon. Iztuzu is a peninsula, a part of which is covered by the sea and the other part is surrounded by the rivers. The coast formed its present shape in 500 BC. At that time, Dalaman River changed its bed because of the overflow of its former bed.

Iztuzu beach holds an award for nature conservation; it is among the small number of beaches where loggerhead sea turtles (Caretta caretta) feel secure enough to leave their young. The beach can be regarded as a real natural wonder. It's claimed to be unique in the sense that it is has fresh water at one side and the Mediterranean Sea on the other. The beach, starting from the sea towards the Dalyan delta, is covered by fine golden sand that has a velvety feel under your feet. The sea itself is shallow and as pure as crystal. While walking on the beach of Iztuzu, this natural wonder, you will even envy your own shadow.

The Sea Turtles; *Caretta Caretta*

The endangered species of Caretta caretta turtles come and nest in these coastal strips of Iztuzu. Holiday makers love the fine sands of Iztuzu no less than the tortoises.

Caretta Caretta is a species of turtle living in the Mediterranean. It is in danger of becoming extinct in all Mediterranean countries except Anatolia. In Turkey, the breeding region of Caretta Caretta is the coastline between Dalyan and Anamur, and the most important place is the Dalyan-Iztuzu coast.

The Sea turtles have become the symbol of Dalyan, and there is particular interest is their mode of reproduction. First they dig furrows in the sand and deposit their eggs, each turtle laying about one hundred eggs in a given furrow. The eggs are then carefully covered over to guard them from discovery and harm. The intriguing thing is that the eggs are laid so as to hatch at the time of the full moon after an incubation period of 60 days.

The young hatch and seeing the light of the moon shining on the sea they move towards it. At that moment, choosing the right direction to move is essential to their survival, and everything is done at Dalyan to make sure that nothing interferes with them. At Iztuzu it is forbidden to turn on lights in the evening, or even to play music.

Caretta caretta

Mudbaths and Thermals

Visitors who come to Dalyan to spend their vacation here, or even those who just happen to be there for a short time, generally do not miss the opportunity to have a photograph of themselves taken, covered in mud.

Dalyan mudbaths

Apart from its natural beauties, Dalyan is also known as the "healing place". It has been said that even the ancient Queen of Egypt, Cleopatra, frequented the mud baths of Dalyan regularly. These natural baths are one of the most popular places for famous people from around the world.

Where the Dalyan River leaves Lake Köyceğiz there are deposits of mud that soften the skin, smooth wrinkles, and beautify. This "beauty mud" is also said to help in cases of rheumatism, sciatica, and slipped disk. There are hotels near the mud deposits, where visitors stay who wish to have a mud cure, either for beauty or reasons of health. Even those who originally intended simply to tour the Dalyan and Caunos area are drawn to the mud, which is washed off in a pool provided for that purpose.

Dalyan mudbath

This is one of the ten most important thermal springs and mineral spring spas in Turkey. The water of the Sultaniye thermal baths, high on a hill to the southwest of Lake Köyceğiz, has been used since Hellenistic times, first by the Carians then the Byzantines. It has a temperature of 39°C, and radioactivity level 98.3. This is the highest level of radioactivity of all the thermal stations in Turkey. Indeed, from the standpoint of their radon measurement these baths are second in the world. The radioactive element radon is especially important in the water. According to experts, radioactive elements are good for different illnesses. The water has properties that are of benefit in cases of neuralgia, neuritis, rheumatic ailments, skin disorders and calcification. When imbibed it helps cure disorders of the liver, spleen and bowels.

It is impossible to ignore the beauty of Dalyan's unique environment; the beach, lake, river and mountains provide a stunning natural setting for every holiday activity.

Thermal bath

References

Dalyan; Kaunos, Köyceğiz, Dalaman, The Blue Voyage. (1991). The State Ministry of Environmental Concern, p. 12–41.

EGENET Tourism Promotion Limited Company, Netcity Reklam. Fethiye-Muğla, Turkey.

Hengirmen, M. Personal Archieve. Engin Publisihing, Ankara, Turkey.

Turkey Voyager Guide. (1996). Ekin Publishing Center, İstanbul, Turkey, p. 247–251.

Türkiye'nin Antik Kentleri, Gezi'99. (1999). Ekin Publishing Center, İstanbul, Turkey, p. 68–69.

Turkey;Hotels, Marinas, Restaurants, Camp Sites.(1993). The Assoc. of Tourism Investors & Delta Group, İstanbul, Turkey, p. 206–207.

The South Aegean, (1990). The Ministry of Tourism General Directorate of Information, İstanbul, Turkey.

http://web.bsu.edu/istanbul/day8_11.html

http://www.math.umn.edu/~alayont/turkiye/ege/ege.html

http://www.wordtravels.com/Resorts/Turkey/Mediterranean+Coast/Dalyan/Map

http://www.mymarmaris.com/sightseeing/caunos.phtml

http://www.allaboutturkey.com/caunos.htm

http://bluevoyagetour.blogspot.com/2006/10/admiral-blue-voyage-tour.html

http://www.enjoyturkey.com/info/sights/dalyan.htm

http://www.dalyanresort.com/?version=1

Damlatas Cave, Alanya

Ebru Alakavuk and Zeynep Yağmuroğlu

The most important sea caves of Turkey are found on the Alanya Peninsula. One of these caves formed by sea erosion is Damlatas Cave. The cave is situated on the west coast of the peninsula, 100 m from the coast. The entrance to the cave is a few meters above the water and was formed within semi-crystalline limestone of the Permian Period. It was shaped by erosion over thousands of years and discovered accidently in 1948 (Aygen, 1984).

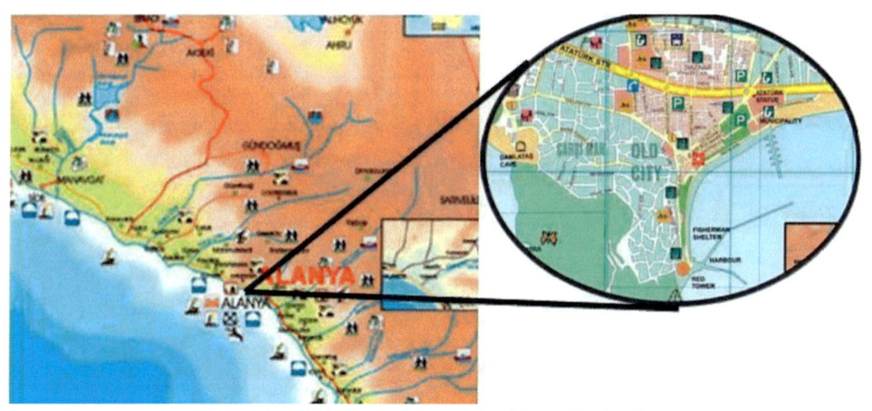

Location of Damlataş Cave on the Alanya Peninsula

Damlatas cave is one of the natural reserves of Turkey because of the morphological structures that are formed in the cave. These morphological structures are rare and that is why the cave is designated as natural conservation zone.

The cave has been formed by many geological factors: tectonics within the metamorphic limestone, faults within these systems, wave action and chemical dissolution by surface water containing large amount of carbon dioxide (Aygen, 1984).

E. Alakavuk (✉)
Yasar University Kazim Dirik Mahallesi 364 Sok, Bornova, İzmir, Turkey
e-mail: ebru.alakavuk@yasar.edu.tr

N. Evelpidou et al. (eds.), *Natural Heritage from East to West*,
DOI 10.1007/978-3-642-01577-9_35,

Damlatas cave takes its name from the dropping stones which are a morphological structure of travertine deposits. "Cave travertine" or "dropping stones" are created by the precipitation of carbonate rocks deposited by the underground water. Cave travertines are developed by the reduction of the pressure of carbon dioxide. The dropping stones made from water containing calcium bicarbonate, which flows through cracks, are named according to their shapes: stalactites, stalagmites, columns, wall dropping stone, curtain-flag dropping stone, cave shields, cave flower, cave pearl and cave stone (Ayaz, 2002).

Stalactites are cylindrical or cone shaped dropping stones formed by water dropping from the cracks in the ceilings of the cave. The growth of stalactites depend on the continuity of physico–chemical properties, morphology and hydrological development of the region and time. Stalagmites are dropping stones found on the floor of the cave, shaped by the mineralized solutions and deposition of calcium carbonate from overhead cracks and stalactites. Columns are shaped by the joining together of stalactites and stalagmites (Ayaz, 2002). Traventines are not valued for industrial use but they are important, for their extraordinary shapes, as tourist sites and natural formations.

Damlataş Cave (Ozansoy, 2006)

Damlatas cave consists of a circular hall with a diameter of 10 m and a height of 15 m, containing stalactites and stalagmites. A corridor 45 m long extends from the hall. The long corridor descends steps to the beach that forms the floor of the cave. Large columns are found in the section from the entrance up to the circular hall (Aygen, 1984).

Dropping stones of Damlataş Cave (Ozansoy, 2006)

Besides its importance for tourists, the cave benefits asthmatics who breathe its air. Scientists analyzed a sample of air to verify the fact that the air was beneficial to patients suffering from non-allergic asthma. They found out that the air in the cave contains 10–12 times more carbon dioxide than normal air and has 90–100% humidity. The temperature in the cave is 22°C. Both the radioactivity and ionization in the cave may contribute to the benefits derived from breathing the air in the cave (Akış, 2007).

Stalactites of Damlataş Cave (Ozansoy, 2006)

Stalactites of Damlatas Cave (Ozansoy, 2006)

There has not been sufficient technical investigation of Damlatas Cave but, because of the extraordinary shape of stalactites and stalagmites, tourism exploded in Alanya after 1987. The number of foreign tourists coming to the area reached 1.464.686 by 2005. These tourists spent 942 $ per person and a total of 1.379.734.210 $ in the area (Akış, 2007).

The Damlatas Cave is not fully open, with tourism facilities. Only a small part of it can be visited. More technical research must be done on the cave in order to fully identify its properties.

References

Ayaz M. E., 2002. Traventenlerde Gözlenen Morfolojik Yapılar ve Tabiat Varlığı Olarak Önemleri. Cumhuriyet Üniversitesi Mühendislik Fakültesi Dergisi.

Akış A., 2007. Alanya'da Turizm ve Turizmin Alanya Ekonomisine Etkisi. Selçuk Üniversitesi Sosyal Bilimler Enstitüsü Dergisi.

Aygen T., Türkiye Magaraları (Turkish Caves), 1984. Turkiye Turing ve Otomobil Kurumu Yayınları.

Ozansoy C., Mengi H., 2006. Mağarabilim ve Mağaracılık. Tübitak Yayınları.

The Dilek Peninsula: Büyük Menderes Delta

Ebru Alakavuk and Burcu Şengün

The Dilek Peninsula (Dilek mountain) is in the northern part of the Büyük Menderes Delta, while marshy fields and lagoons along the Delta coast lie to the south. These two natural units, which seem to conflict with each other, are actually complementary units of the same environment. The lower level of the Büyük Menderes delta to the south balances the uplands of Dilek Mountain to the north.

Turkey

The National park, including the Samsun Peninsula, is in Güzelçamlı district of Kuşadası. This district is one of 3 places that has had an "environmental award" presented by the European Parliament. The national park protects a area of 11,000 ha and one third of it is public. The remaining part of the park is a preserve for wild animals (Sezer, 2006).

E. Alakavuk (✉)
Yasar University Kazim Dirik Mahallesi 364 Sok, Bornova, İzmir, Turkey
e-mail: ebru.alakavuk@yasar.edu.tr

N. Evelpidou et al. (eds.), *Natural Heritage from East to West*,
DOI 10.1007/978-3-642-01577-9_36,

Büyük Menderes Delta National Park covers an area of 27,675 ha in the area of Aydın. 10,985 ha of this area belong to Dilek Peninsula which became a national park in 1966. The Büyük Menderes Delta covers the remaining area of 16,960 ha and was added to the national park in 1994 (Sezer, 2006). The Dilek Peninsula with a length of 20 km and an average width of 6 km is formed by Samsun Mountain projecting into the Aegean Sea. The geological structure of this peninsula consists of paleozic, mezozoic limestones, marbles and neogenic sediments. The peninsula has an attractive coast with its sandy and clayey beaches. It has numerous hills, valleys, canyons and bays (www.dilekyarimadasi.gov.tr).

The Dilek Peninsula- Büvük Menderes Delta (Aydin Çevre Durum Raporu, 2006)

After human communities became settled, technological developments began to harm the natural environment. Forested areas were reduced; air and water sources were polluted. Now, as a result of the decrease in agricultural fields, some species of flora have become extinct.

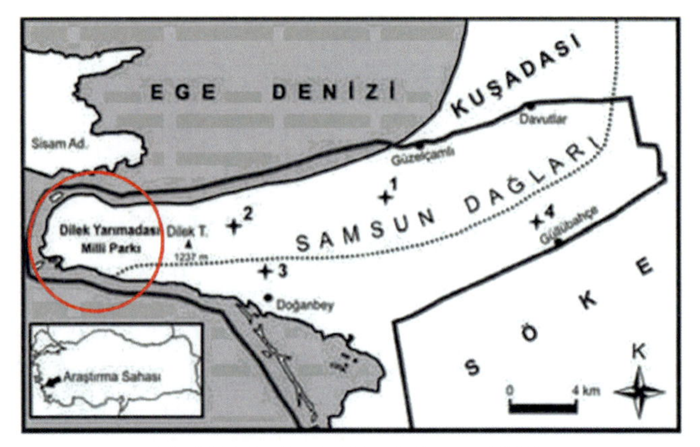

The Dilek Peninsula (Koç & Yağmur 2007)

The idea of a "National Park" is crucial to the protected areas concept which aims to preserve natural resources. A National Park is a protected and tourist area which has rare natural and cultural sources, and also a scientific and an aesthetic dimension. At the present time it is important to preserve biological variety because of global ecological problems. Areas which have biological diversity are turned into National Parks for his purpose. Turkey has 33. Dilek Peninsula-Büyük Menderes Delta national park is an important natural resource of Turkey and the world. The national park contains Mediterranean and European Siberian flora of international importance (Sezer, 2006).

Viburnum tinus (www.wikipedia.org)

The varied physical properties of the Dilek Peninsula and the Büyük Menderes Delta have produced a wide variety of vegetation within a small area. In this national park the number of identified plant species is 804, belonging to 95 genera. Six of these plants can only be seen here. Besides this, there are 31 plants endemic to Turkey. Examples of almost all kinds of Akdeniz Lemur Vegetation is alive and healthy here. This region is the only place where small forest groups are formed by *Castanea sativa*, Sage tea, *Viburnum tinus, Cupressus sempervirens, C. horizantalis, Juniperus phoenicia, Quercus ilex* & *Q. coccifera*. These groups of plants growing between the Mediterranean and the Black Sea can be seen together only in this

National Park. As for fauna, 36 kinds of mammals, 42 kinds of reptiles, 45 kinds of marine animals can be found on Dilek Yarımadası. Besides dolphins and sea turtles, cephalopods, sea urchins, starfish, sponges and many kinds of fishes live together here. Whales also can occasionally be seen on the coast of the national park, in 1998 a 14 m long Fin Whale, *Balaenoptera Physalus*, indigenous to the Mediterranean, was found in Kavaklıburun Bay. Beyond the easternmost end of the peninsula, the endangered leopard, *Panthera pardus tulliana* lives. *Monachus monachus, Sus scrofa*, Caracal, Lynx, *Canis aureus* and Hyaena are some of the animals that live in this zone (Sezer, 2006).

In 1966, the Dilek Peninsula-Büyük Menderes Delta was made a natural park because of the *Panthera pardus tulliana* living in this region. In 1994, the coast of Büyük Menderes Deltası was also made a natural park because of the rich population of birds whose increasing populations depend on the varied physical properties of the area: forest, wetlands and mud flats (www.sirince-evleri.com).

The Dilek Peninsula (Aydin Çevre Durum Raporu, 2006)

The 584 km long Büyük Menderes River, running in the southern part of the park, is the most important water source of the peninsula. The rapid evolution of the mouth of the delta has produced many lagoons, saline bogs and mud flats. This region has international importance because of its biodiversity, including endemic as well as endangered species. It is protected by the Ramsar, Bern and Rio Agreements and also by the Barcelona Convention (Sezer, 2006).

Natural Parks are the most important examples of nature conservation; however, successful nature conservation must be supported by the work of people who are experts in their fields.

References

Sezer, İ., (2006). Dilek Peninsula (Büyük Menderes Deltası Milli Parkının Coğrafi Etüdü, AYDIN ÇEVRE DURUM RAPORU), Master Thesis, Atatürk University, Erzurum.

KoçH.&Yağmur E. A., (2007). Dilek Peninsula National Park (Söke-Kuşadası, Aydın) Akrep Faunası, Journal of Ecology: 52–59, İzmir.

www.dilekyarimadasi.gov.tr

www.sirince-evleri.com

wowturkey.com/forum/viewtopic.php?t=6008

en.wikipedia.org/wiki/Image:Viburnum_tinus.JPG

www.ubcbotanicalgarden.org/potd/castanea_sativa1.jpg)

www.turkeyforum.com

Halfeti – Rumkale

Gülnur Ballice

In Halfeti, steppe life prevails framed by natural living of rocks and water. You can observe animals living in the steppe, rocks and river. The Euphrates flows along the rocks bordering Halfeti. You can see rare animals living harmoniously in the environment of rocks and water. Most of Halfeti has been submerged in water by the Birecik Dam, which was built in 2001 between Birecik and Halfeti. Consequently, many inhabitants of Halfeti had to migrate or move to new settlements.

Satellite photo of Halfeti and Rumkale
(http://www.sitemynet.com/ eskihalfeti/)

G. Ballice (✉)
Yasar University İzmir, 35500, Turkey
e-mail: gulnur.ballice@yasar.edu.tr

N. Evelpidou et al. (eds.), *Natural Heritage from East to West*,
DOI 10.1007/978-3-642-01577-9_37,

Location of Halfeti (http://upload.wikimedia.org/wikimedia)

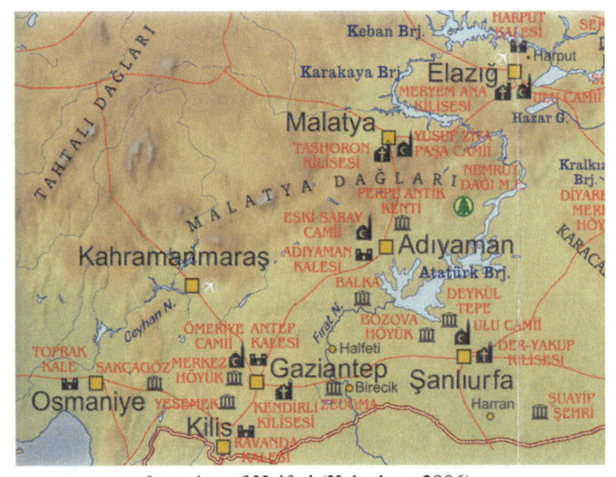

Location of Halfeti (Kabadayı, 2006)

Halfeti is a district center on the east bank of the Euphrates and is part of the province of Şanlıurfa. During the Ottoman period it was the center of Rumkale township, which was located in the Urfa subdivision of the province of Aleppo. At the beginning of the twentieth century there was one government guest house in the village, 16 mosques and smaller places of worship, two libraries and schools, 55 stores, four bakeries, 48 mills, one bath and one tannery. Moreover, it is known that there were more than 1000 gardens and nearly 3000 orchards in the township.

The ancient city of Arulis and stone quarries are very close to the village (Başgelen, 2000)

Değirmendere
(http://www.sitemynet.com/ eskihalfeti/)

Entrance to the Değirmendere
(http://www.sitemynet.com/ eskihalfeti/)

Değirmendere
(http://www.halfeti.gov.tr)

Gümüşyaka village
(http://www.galfeti.gov.tr)

The Black Rose is grown only in the town of Şanlıurfa , Halfeti. Once it is moved to another place, it changes color to red. The Halfeti Roses that change from dark red to black color in time, are small, multilayered and scented. The plants are about 1–1.5 m tall and have flowers with diameters 6–7 cm.

Rarely found animals and plants living harmoniously in the environment of rocks and water (Doğa Derneği leaflet)

Black Rose flourishing only in Halfeti in Turkey (http://www.sitemynet.com/ eskihalfeti/)

The houses of Halfeti, set in winding streets, form part of the landscape. Generally two, sometimes three, storey, all these houses built of white stone are adorned with elaborate stencil work. A captain living in Halfeti has described the houses as follows: "The houses of Halfeti have bird's nests on their roofs. Each one has a view of the Euphrates, and no house obstructs the view of the others. Each one has a garden where in the old days the black roses unique to this part of the world were grown. To cool off the hot weather, wooden platforms known as 'taht' were brought up to the balconies and people spent the night under the stars.." (Pekdemir, 2007).

The last remaining Halfeti houses (Pekdemir, 2007)

The Bey Mansion built at the beginning of the 20th century (Başgelen, 2000)

The historical structures in the village include the "Bath House" dating from 1867 to 1868, the historical Latifzade Bath from the period 1796 to 97, the Central Mosque built in 1804 and repaired in 1906, the Bey Mansion dating to 1910–1913, and the Muhittin Ganneci house. In Cekem village, another area of the district which will be completely submerged, there is a building complex which consists of a small mosque, tomb and enclosed mausoleum and an old cemetery with interesting tombstones.

The Bey (Emir) Mansion, looking northeast (Durukan, 2003)

The Bey (Emir) Mansion, detail (http://www.sitemynet.com/ eskihalfeti/)

Latifzade Bath, dating to 1796–97, and the central Mosque as the dam's water begins to rise (Başgelen, 2000)

The Latifzade Bath and the guesthouse behind it before the water level rose (Başgelen, 2000)

During the surface investigation which G.Algaze made in the Euphrates Valley (starting at Halfeti), it was determined that there were three settlements in the vicinity of Halfeti, including the mound with an ancient cemetery on top of it

The Central Mosque, Halfeti (http://www.sitemynet.com/eskihalfeti/)

The Central Mosque, Halfeti (http://www.sitemynet.com/eskihalfeti/)

The Halfeti Mosque, (Doğa Derneği leaflet)

A last look at the Halfeti shore from Kalemeydanı (Başgelen, 2000)

The famous Feyzullah Efendi Mansion in Kalemeydanı while being torn down before the water rose (Başgelen, 2000)

A view of Kalemeydanı and Halfeti in 1997 (Başgelen, 2000)

Stone mansions at Gümüşgun (Ebnes) village on the west bank of the Euphrates, which is connected to Gaziantep, and vaulted antechambers with a view of river (Başgelen, 2000)

Stone mansions at Gümüşgun (Ebnes) village on the west bank of the Euphrates (Başgelen, 2000)

at Kalemeydanı on the Euphrates west bank. Settlements were also uncovered at Karacali and Ören Bahçe.

Sources say that up to the end of the nineteenth century, Halfeti was surrounded by forests. At the begining of the twentieth century, it supplied wood and coal to Antep, Birecik and Urfa. Administratively tied to Gaziantep until 1924, it became an administrative subdivision attached to the district of Birecik in 1926. In 1954 it became a district center in the province of Urfa.

Rumkale 1880 (Başgelen, 2000)

RUMKALE

Rumkale is located to the north of the Birecik plain, lying along the east bank of the Euphrates above a hill looking to the Şanlıurfa road. It borders Birecik on its north and northeast (Sachau, 1883). The capital of the province was Halfeti. Its site accordingly comes within the province of Gaziantep. The west part of the province is rocky, whereas its eastern part is smooth and fertile.

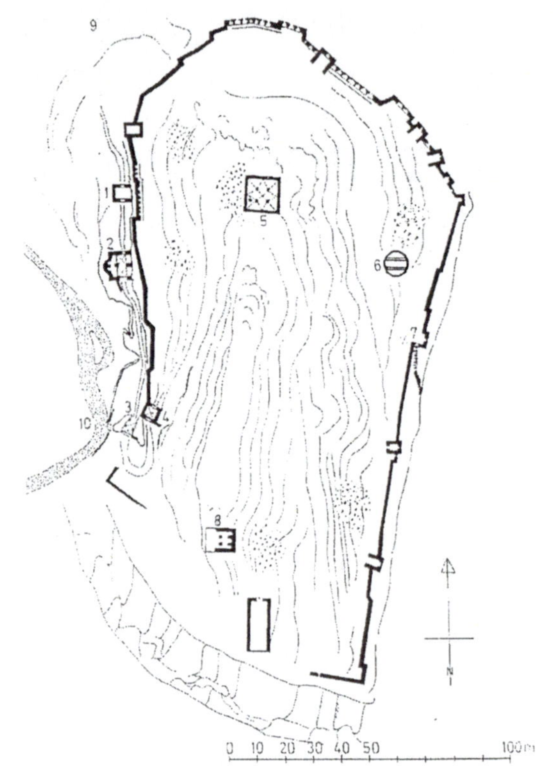

Plan of Rumkale, Hellenkemper 1976 (Durukan, 2003)

A view of Rumkale (Başgelen, 2000)

The arsenal at the foot of the fortress (Başgelen, 2000)

Sixty-two kilometers northeast of Gaziantep there are cliffs where the Merzimen Stream flows into the Euphrates. This strategic area is thought to have been the site of Shitamrat conquered by the Assyrian King Salmanassar III in 855 B.C. The fortress benefited greatly from the topography of the steep cliff on which it was built, rising straight up from the shores of the Euphrates and the Merzimen (Başgelen, 2000).

Water well in the fortress, going down to the level of the Euphrates, 8 meters wide and 75 meters deep (http://www.sitemynet.com/eskihalfeti/)

Some of the ruins in the fortress (http://www.sitemynet.com/eski halfeti)

Rumkale, looking east (Durukan, 2003)

Among the architectural works surrounding it, including the ruins of the St. Nerses Church and the Bersauma Monastery, is a remarkable water system composed of a wide, cylindrical well for air-conditioning and a spiral road from the corner of this well down to the level of the Euphrates. It is said that the Disciple John copied and hid the manuscript of the Gospels here while he was in seclusion; later on the copies were taken to Beirut. Known as Hromgla, this area was an important center in the eleventh century when the Crusaders ruled Urfa. In the middle Ages the fortress-city was called "Hromklay" by the Armenians, Kalarhomate by the Syrian Christians, and Ranculat by Westerners. It passed into the hands of the Mamluks at the end of the twelth century, who first called it Kal'at et Muslimin. In the thirteenth century the Jacobite Patriarch Ignace II had a church built out of the rocks at Rumkale. The area passed into the hands of the Ottomans after the Battle of Marj Dabik, and it became a township attached to the Birecik Sancak of the province of Aleppo.

Different views of Rumkale (http://www.halfeti.gov.tr)

In Ottoman times it was known in the sixteenth century as the Kale-i Zerrin or Golden Fortress (Pekdemir, 2007).

Views of Rumkale from the eastern bank of the Euphrates in 1996 (Başgelen, 2000)

Views of Rumkale from the eastern bank of the Euphrates when the water level had risen in May 2000 (Başgelen, 2000)

Rumkale, looking south (Durukan, 2003)

Evliya Çelebi, who visited Rumkale in the middle of the seventeenth century, wrote that when Sultan Selim captured the fortress from the Mamluks in 1516 he tried to improve it. The fortress included a mosque, bath and small market. Richard

Pococke, an eighteenth century traveler, wrote that there were a number of magnificient buildings on the hill and a small but very beautiful church in the Gothic style. As for Moltke, who came to Rumkale in April 1838, he said, “It is difficult to say where the rocks and the work of man begins”. Karl Humann, who visited the site in May 1883, saw a “richly crowned, heap of columns” on the hill. In the Euphrates Valley between Samsat and Rumkale there are many places which have been carved out of the rocks. When the Birecik Dam is completed, Rumkale will become a peninsula (Başgelen, 2000).

References

Kabadayı, E. (September 2006). Güneydoğu, Atlas Seyahat Kitaplığı, İstanbul: Doğan Burda Yayıncılık ve Pazarlama A.Ş.

Başgelen N., Ergeç, R. (2000). Belkıs/Zeugma, Halfeti, Rumkale, İstanbul: Archaelogy and Art Publications, TOFAŞ with the support and assistance Türk Otomobil Fabrikası A.Ş.

Doğa Derneği Booklet, Güneydoğu'nun Cenneti Halfeti.

Durukan, A. (2003). The Cultural Heritage in the towns of Birecik, Halfeti, Suruç, Bozova and Rumkale, Ankara: Republic of Turkey Prime Ministry Southeastern Anatolia Project Regional Development Administration, p. 224–267, 325–335.

Pekdemir, N., Aybudak, K. (December 2007). Going with the flow of the Tigris and the Euphrates, Skylife, p. 53–66.

Sachau, E. (1883). Reisen in Syrien and Mesopotamien, Leipzig.

www.radikal.com.tr/veriler/2007/07/17/gul.gif, 'Güllerin efendisi' Halfeti'ye dönüyor 17/07/2007.

http://upload.wikimedia.org/wikipedia/commons/thumb/e/eb/Tu-map.png/500px-Tu-map.png.

http://www.lavinya.net/galeri/data/media/10/Siyah_Gl.jpg.

http://www.meleklermekani.com/adim-adim-turkiye/54133-siyah-gul-diyari-halfeti.html.

Kekova

Ebru Aydeniz

The coastal area and small hills between Kale and Kaş in the south of Anatolia, west of Antalya, is the Kekova region. There are irregular inlets, peninsulas and about 30 large and small islands in this region. The flora is typical Mediterranean. The heavy smell here has not been destroyed in forest at various points. The inner parts from the coast get higher and they form the fringe of The West Taurus. The coastal morphology is generally rocky, but there are also pebble beaches at various places (Özel Doğa Alanları, 2008).

View from the Karalos inlet (Türkiye'nin Özel Doğa Alanları, 208)

E. Aydeniz (✉)
Yasar University Kazim Dirik Mahallesi 364 Sokak, Bornova, Izmir, Turkey
e-mail: ebru.aydeniz@yasar.edu.tr

N. Evelpidou et al. (eds.), *Natural Heritage from East to West*,
DOI 10.1007/978-3-642-01577-9_38, © Springer-Verlag Berlin Heidelberg 2010

This area is covered by *Campanula lycica* which supports its 'Special Natural Area' status. This species is in danger worldwide. The common shag (*Phalacrocorax aristotelis desmarestii*) and red falcon (*Buteo rufinus*) do not have the same status but are seen here. Mediterranean seals (*Monachus monachus*), also in danger worldwide, may also be seen. This region is an important habitat and potential reproduction area for these animals. Sualtı Araştırmaları Derneği Akdeniz Foku Grubu (SAD-AFAG) control the coast either by boat or on foot and observe them throughout the year. The otter (*Lutra lutra*) also lives here. *Lyciasalamandra luschani* lives in only a few places all over the world, and this region is one of them. *Ophiomorus punctatissimus*, a type of lizard, is an important reptile in the region. A type of butterfly, *Glaucopsyche alexis*, is another important creature living in the area (ÖDA, 2008).

From Kale Village To Kekova island (Türkiye'nin Özel Doğa Alanları, 209)

Kekova island is 500 m from the coast. It is a narrow island 7.4 km long. It is covered partially with pine, olive and carob trees. There is a drinking water spring on the island. Because of these natural beauties there are many ancient sites. Kekova, Simena and Teimiussa faced each other around a lake-like bay. Kekova island sank a few meters following an earthquake in the second century BC. A Keova site on the north coast was submerged. The people who survived the earthquake moved to Simena and Teimiussa, across the island.

Monachus monachus in Kekova region

Kekova region

This region still has important natural and cultural assets, with its underwater ancient sites, natural beauties, ship wrecks today. This area is the least disturbed area in Turkey. There are only a few settlements. Density of population is very low, but the transport conditions are good.

There are many archeological sites on the island. The port walls, stores, streets and stone stairs of the sunken Kekova site at the north shores are very attractive. Some parts of the houses are drowned whereas the rest are above water. The apse of the Byzantine church is by the Tersane inlet; the church belongs to orthodox missionaries, and has a square plan and frescoes from the fifth century.

An aerial view of Kekova Island

A general view of Kekova Island

Kekova Island-Tersane Cove. The church apse found here collapsed in recent times

A view of the sunken city of Kekova

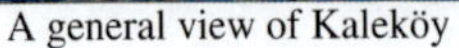

A general view of Kaleköy

The ancient city of Simena, located in Kaleköy, has become a part of modern life

Ancient city wall of Simena

The ruins of the Simena site across Kekova island is located around Kale village at present. This site retains traces of various civilisations. A Medieval castle on top of the hill is in good condition. The masonry styles of different periods are reflected on these castle walls. Fragments of ancient ruins may be seen to the south of the castle. A small theatre for 400 people is in the castle. The 8 rows of seat are carved in the rocks. The stoa of The Poseidon Temple is outside the walls facing the sea. A ruined bath constructed for Emperor Titus is by the sea shore. The Necropolis is at the north, in an area of the house ruins.

The ruins of Teimiussa site are in Üçağız village. The necropolis area with Hellenistic and Roman sarcophagi is to the east of the village harbour. A small village with rectangular walls is by the sea. The walls were enlarged during the Byzantine period and surrounded the whole site.

A general view of the Üçağız

The necropolis of the ancient city of Theimussa, situated in Üçağız

A house-type Lycian tomb which belonged to Kluwainimi, found in Üçağız

Ancient Theatre in fortress

Poseidon Temple remains, outside of the city

References

Aksit, I., (2000). The land of light Lycia. Istanbul: Aksit Kultur ve Turizm Ticaret Ltd. Sti. 108–119.
Bayburtluoğlu, C., (2004). Lykia. Istanbul: Suna-Inan Kıraç Akdeniz Medeniyetleri Arastirma Enstitusu, 225–234.
Kunar, S., (2000). Istanbul: Net Turistik Yayınları A.Ş., 29–42.
The church of St. Nicolas in Myra (1998). Ankara: Antalya Museum Publication IV, 78–101.
Türkiye'nin Önemli Doğa Alanları (2007). Ankara: Doğa Derneği, 208–209.
Umar, B., (1999). Lykia: bir tarihsel coğrafya araştırması ve gezi rehberi. İstanbul: Inkilap, 104–112.

Mt. Nemrud (Nemrut) Kommagene (Commagene)

Malike Özsoy

Adıyaman, the cradle of the oldest civilizations in history, is among the most important provinces in Turkey. Especially, on the Nemrut Mountain in Kahta District, the graves, temples and the statues of kings are extremely interesting. The Nemrut Mountain National Park and the summit of Nemrut Mountain is a charming place in the province with its impressive silhouette, natural beauty and historical assets. It has a unique pastoral beauty, where the sunrises and sunsets are the best in the world. This place is considered the eighth wonder of the world, a heavenly throne of the gods, the world's highest open air museum. The statues themselves are impressive, but the view is the most spectacular, especially in early spring when the plains are full of long grass speckled with the occasional red poppy.

The Situation of Mt Nemrud on the Map of Turkey

M. Özsoy (✉)
Architecture Department Yasar University, 35500 Bornova-İZMİR
e-mail: malike.ozsoy@yasar.edu.tr

N. Evelpidou et al. (eds.), *Natural Heritage from East to West*,
DOI 10.1007/978-3-642-01577-9_39, © Springer-Verlag Berlin Heidelberg 2010

Mount Nemrud is a dormant volcano in Eastern Turkey. It lies within the boundaries of Karadut village, Kahta District and Adıyaman Province. It is 54 km to the north of Kahta, 100 km to Adıyaman. Adıyaman is a province which covers most of the Commagene lands. It is located in a valley on the outskirts of Southeastern Taurus. The peaks of the Antitaurus range bordering the northern part of the province are Mt Akdağ, Ulubaba, Recep and Nemrud (2,150 m) (Gökovalı, 1988).

View of the Taurus mountains from Mt Nemrud

The Temple of Gods on Nemrut

Mount Nemrud is home to the "Temple of Gods"of Commagene and the hierothesion (tomb sanctuary). Egemen Sarıkaya says "Unless deliberately shaped, mountains do not result in summits or peaks such as this". In fact there is an artificial mound situated on the natural summit of Nemrud. This artificial mound is the tumulus of the first century BC Commagene king, Antiochus I Epiphanes, one of the kings of Commagene (Sarıkaya, 1982). Even today, despite losing a third of its original height over the centuries, the massive tumulus built to house Antiochus' body can be seen from over 100 km away. It has even made the Guinness Book of Records as the world's largest man-made mound, a shade under 60 m high and covering an area of 7.5 acres (USA Turkish Times). This tumulus and the thrones of gods on the eastern and western terraces and the other works are all together called the "Hierothesion of Antiochus I" or "temple of gods". The tumulus may be seen from anywhere in the kingdom; and this must have been the reason why Atiochus I chose it for his Hierotheison. On the other hand, the belief "the higher the place, the nearer the gods" which had been upheld by man for centuries, could have been another reason. The king wanted religious ceremonies to be held here between the dates of his birth, and his ascension to the throne, 10-Loos (July), (Gökovalı, 1988).

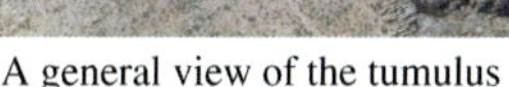

A general view of the tumulus

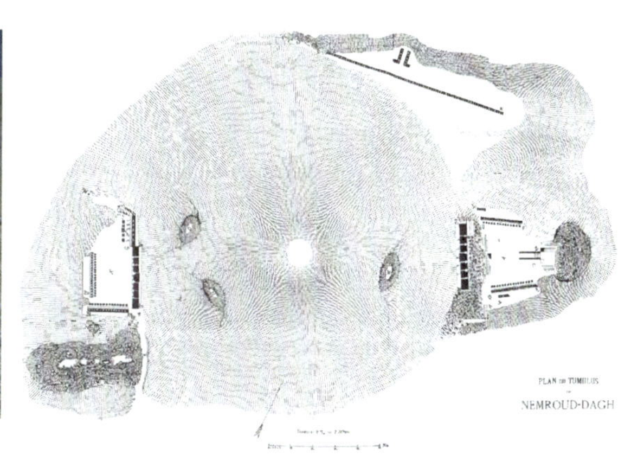

The Plan of Tumulus

Antiochus I, like his father, Mithradates Callinicos, spread Hellenic Culture and language in his country and fused Hellenic Culture and Persian religion together. He associated himself with the Persian Emperor Darius on his father's side and with Alexander the great on his mother Queen Laodike's side. In this way, Commagene kept good relations with the Romans in the west and with the Persians in the east. These giant monuments not only show the magnificence of the Commagene civilization but how the Hellenic and Persian influences had melted in the Anatolian crucible, (Gökovalı, 1988). Befitting a tiny kingdom straddling the ancient borders between the Greek–Roman world of the west and the oriental Persian world to the east, the gods portrayed are a fusion of the two civilizations, (USA Turkish Times).

The mausoleum of Antiochus I is surrounded by three sacred areas in the shape of a terrace carved into the hard rock, to the east, west and north. Limestone was used in laying the terrace walls supporting walls, and in making the statues of idols and

View of the rocks forming the statues of Gods

View of the rocks forming the statues of Gods

thrones. The reliefs and the pedestals they stood on, and the lion statues, were carved from green sandstone. Most of the material was cut from the calcareous rocks and some of it was brought from the Gerger Castle and Karabelen rock mines, (Önen, 1987). How these rocks weighing tons were carried from kilometers away, 2,000 years ago, using mules to climb Mt. Nemrut, is a question to be borne in mind.

The ascending ceremonial road approaches the tumulus from the south. On the east and west terraces, there are symmetrical reliefs, eagle and lion statues, in addition to the five thrones on each side surmounted by statues. The height of the statues is close to 9 m with their thrones.

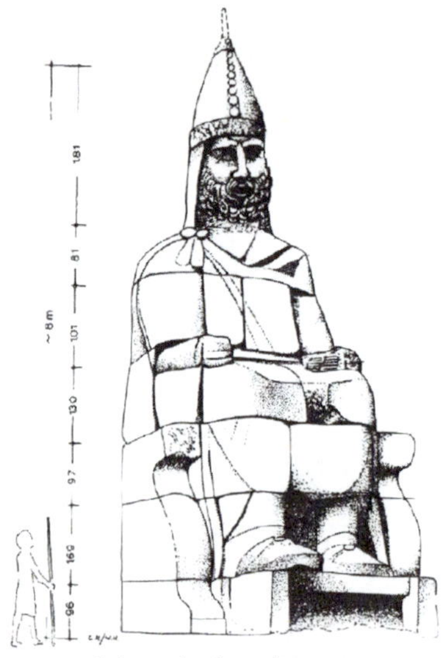

Schematic view of statues

View of statues

On the east and west terraces there are numerous lion and eagle reliefs and statues. As is known the lion has symbolized power and sovereignty. The eagle is supposed to be the nearest to god as it can fly high and is the sacred animal of Zeus, the god of gods.

Lion statue on Nemrut tumulus

Eagle statue on Nemrut tumulus

East Terrace

At the eastern terrace are located the statues of Apollo Helios Hermes, the god of art; Tyche (Fortuna) Kommagene, the goddess of fertility and fortune; Zeus Oromasdes, the god of the heavens; Herakles Artagnes Ares, the god of strength; King Antiochus I; an eagle and a lion. Antiochus believed himself to be the descendent of Apollo so he built a statue of himself along with those of Apollo, Zeus, Tyche (Fortuna) and Herakles flanked by a lion and an eagle. (Başgelen, 2003) The steles of the Commagene Royal Family are to the north and south; and to the east of the terrace, there is a rectangular shaped altar with steps, and beside it a protective lion statue. The area of the pyramidal fire altar is roughly 100 m^2 (Gökovalı,1988)

- King Antiokhos 1
- Tyche, Kommagene
- Zeus, Oromasdes
- Apollon, Helios Hermes
- Herakles, Artagnes Ares

East Terrace with five statues of Gods

East Terrace Animation: five statues of Gods, eagle, lion and altar

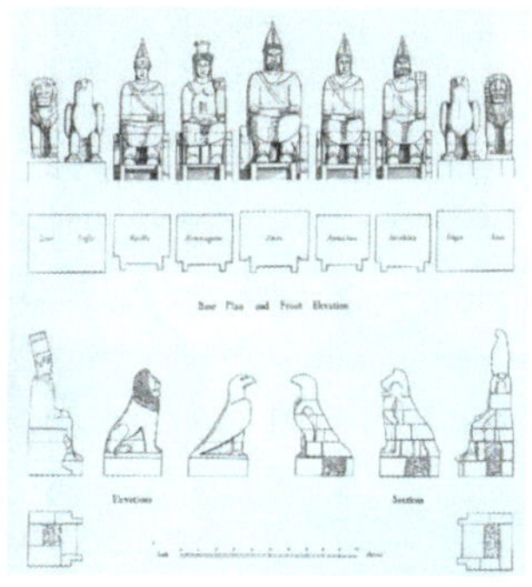

Schematic view of five statues of Gods, lion and eagle

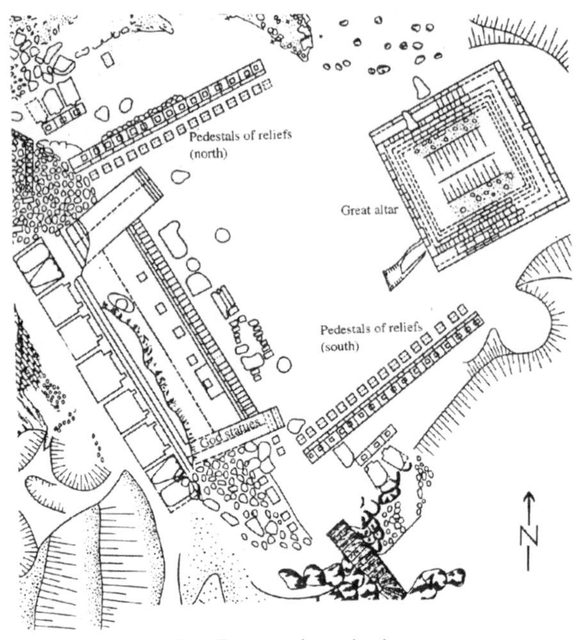

East Terrace schematic plan

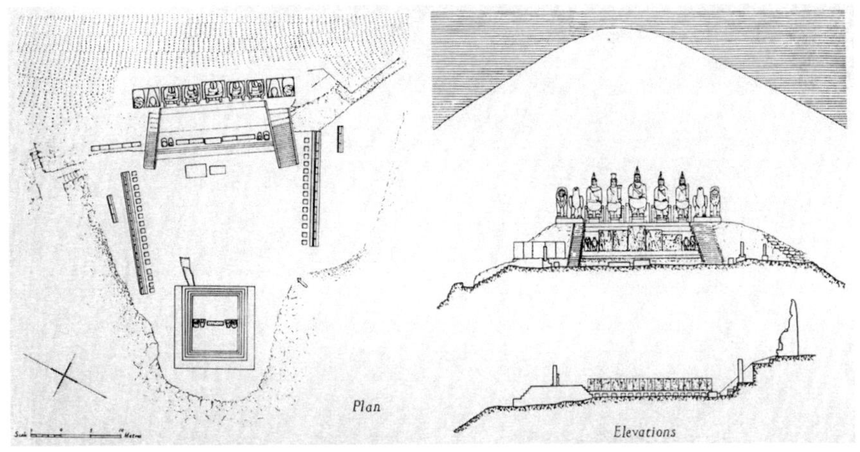

East Terrace reconstruction (drawn by G.R.H.Wright, 1955)

Western Terrace

The western terrace, where there are the same statues, is more effective in its sculpture, in spite of the fact that it has experienced more damage in comparison with the eastern terrace.

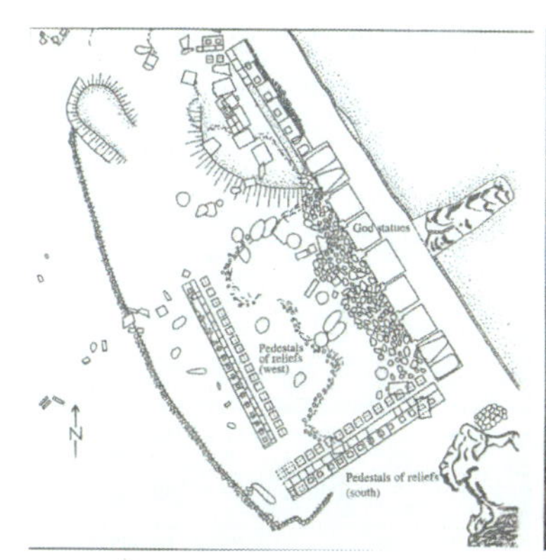

West terrace schematic plan

View of west terrace

The row of statues on the western terrace depicts the same deities and there are also reliefs showing King Antiochus shaking hands with Apollo-Mithra-Hermes, with Zeus-Oromasdes, and with Herakles-Ares-Artagnes (Cimok, 1995).

Animation of the Western Terrace reliefs

King Antiochus shakes hand with Zeus

Among the carved stones on the western terrace, one known as the "Lion of Commagene" bears significant astronomical information. Nineteen stars may be seen in the background of the relief and on the lion's body, a crescent moon is shown on the lion's neck, and above the lion's back are three planets named Mars, Mercury and Jupiter. This relief has been interpreted (by archaeoastronomers using the Skyglobe computer program) to indicate a date of July 6th on either 61 or 62 BC. Different opinions exist as to the significance of this date. Professor Otto Neugebauer of Brown University believes it is the date when Antiochus was set on the throne by the Roman general Pompey, while Adrian Gilbert (in *Signs in the Sky*) sees it as an esoteric coronation of Antiochus as head of a secret Persian / Anatolian brotherhood (Gray, 2007).

Zodiac Line

Mount Nemrut has a unique pastoral beauty, especially at sunset on the western terrace, and visitors experience moments that they will not forget as long as they live. The most suitable time of year for climbing the mountain is between 15 May and 15 October.

The Statues on the Thrones

The statues on the thrones are protected by lions and eagles; four are male, one is female. Here is some information about their identities and history:

1-Antiochus 1, Epiphanes

The great King, son of the Mithradates Calinicos – one of the great Commagene kings- and Queen Laodike, considered himself as god. The height of Antiochus head is over 2.5 m, that is to say it is the second biggest amongst the heads in the temple of gods after that of Zeus.

2-Fortuna, Commagene

This is the only female god (goddess) on the west and east terraces. Her name in Commagene tradition is not known. This is the goddess of fortune and Fortuna, the Roman goddess of fortune, as a fertility goddess. On the head of the dignified goddess are seen vegetable and fruit carvings with grape and pomegranate reliefs between them.

3-Zeus, Oromasdes

This is a god which unifies Zeus, the chief god of Hellenism, with Oromasdes (Ahura Mazda), the chief god of the Persians. Zeus is known as the chief god of classical mythology, (Gökovalı, 1988). His statue is not only the biggest Zeus statue on Nemrut, but also on earth. The monuments were made big enough to be seen by those high above, (Sarıkaya, 1982). These colossal stone blocks were brought from elsewhere, although it is difficult to understand how these heavy stones could be lifted and erected.

West terrace, lion, eagle and head of Antiochus I

West terrace: head of Commagene

West terrace: head of Zeus

4-Apollon, Helios Hermes

Apollon is an Anatolian Lycian god. He is the son of Zeus and the goddess of night Leto. Since the sun was born of night, Apollon is the god of light and sun.

5-Herakles, Artagnes Ares

The real name of Heracles, known to the Romans as Hercules, was Alides. According to Mythology, Zeus created him “for the protection and comfort of gods and humans alike”.

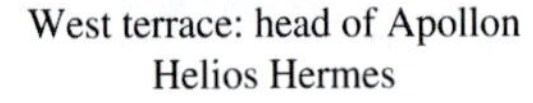

West terrace: head of Apollon Helios Hermes

West terrace: head of Herakles, Artagnes Ares

North Terrace

There is a 10 m wide ceremonial road connecting the west and east terraces. There are, over a distance of 80 m, uncompleted pedestals. This terrace does not have any statues. It is thought that it was used as a place of assembly during the cult ceremonies. The remains of a wall, 80 m long, the stone sockets and the slabs of sandstone have given scholars the impression that this terrace was intended to be used for statues representing the kings who would come to the Commagenian throne after Antiochus I, (Cimok, 1995).

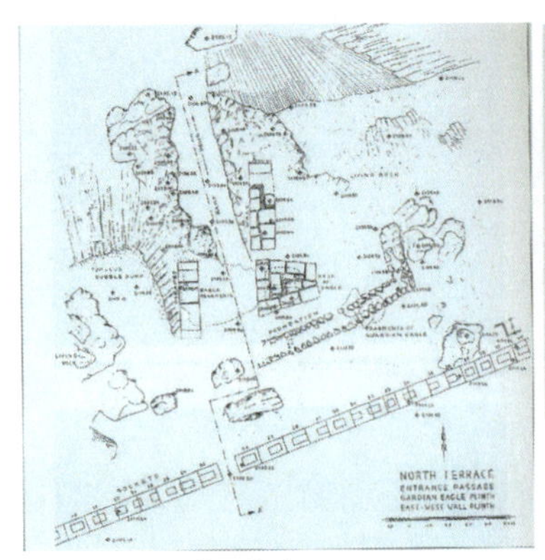

Eighty five Steles on west terrace

North Terrace Plan

The pilgrims assembled at the foot of the mountain from the surrounding valleys. Here they were provided with food and drink by the servants of the priests. From there, two processional ways led to the sanctuary on the mountain. Both processional ways are marked with a stele close to the sanctuary. On these steles is carved a text. Here, Antiochus informs the visitors that they have set foot on consecrated ground and should behave themselves as such. The southern processional way was for the nobles of Kommagene and ended on the West Terrace. The northern was for the common people and led to the North Terrace. At the North Terrace, in the forecourt of the sanctuary, the people were prepared for their meeting with the gods. With some difficulty you can find the worn ramp, where the people entered the North Terrace. From there, they moved in procession to the East Terrace along the 80 m long row of steles, which separates the North Terrace from the rest of the sanctuary. These steles bear neither portraits nor inscriptions, as Antiochus intended them for his descendants. (http://web.deu.edu.tr/atiksu/ana37/ana37.html)

Karakus Tumulus

About 10 km on the road to Nemrud on the west bank of Kahta stream (the Nymphaios), is a tumulus called "Karakus" or "Black bird". The site takes its name from an eagle statue surmounting a doric column on its southern side.

Karakus statue

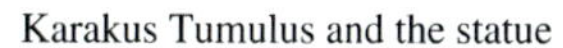

Karakus Tumulus and the statue

The tumulus is the burial ground for the women of the Commagene Royal Family. The Monumental Tomb was built by Commagene's King Mithradates II and was dedicated to his mother Isas. On the other hand, according to some other sources this mound was made for the conducting of rituals during the winter months when ascent to the tumulus on Mount Nemrud would be impossible. The sarcophagus room is covered with cut stones and pebbles. The pebbles on top of the cut stones were carried in from the Kahta stream bed. Thus the piled hill has reached a height of about 20 m while the tumulus has a diameter of 100 m. (Dörner et al., 1981). The eastern, southern and western sides of the tumulus were each decorated with a set of three doric columns. Four of these are still upright, one of which is surmounted by a 2.45 m high statue of an eagle. Of the two columns standing side by side, one of them is decorated with a headless statue of a lion. A ruined lion statue lies on the ground. While on another column a dexiosis, a relief of the king shaking hands with a god, can be seen. The doric columns surrounding the Karakus tumulus are about 9 m high including the reliefs and statues. Not taking the bases, capitals and the surmounting statues or reliefs into account, each column consists of six marble blocks, (Gökovalı, 1988).

Commagene had an art tradition which was completely its own. It was an unique synthesis of Greek and Persian art. Antiochus I stimulated art in a special way. He gathered together at his court a group of artists and scientists. They were called Philoi, the "Friends of the King".

Karakus Tumulus - tomb of the kings of Commagene-
Doric columns

Under the reign of King Mithridates art was still dominated by eastern influences. During the reign of Antiochus, the style became more naturalistic and less stylised. Antiochus himself, preferred Greek culture. He called himself literally a "Friend of Greeks and Romans".

The statues on top of Mount Nemrud became the crowning glory of Commagenian art. Here, east and west fused into total harmony. Any superfluous detail that could possibly disturb the form of the statues has been avoided. There are no luxuriant beards, jewelry and other ornaments. Even today the gazing heads of statues impresses people by their timeless beauty set against spectacular views of Mt Nemrud.

References

Basgelen, N. (1998). *Tanrılar Dağı Nemrut*. İstanbul: Arkeoloji ve Sanat.
Basgelen, N. (2003). *Nemrut Dağı, Keşfi, Kazıları, Anıtları*. İstanbul: Arkeoloji ve Sanat.
Cimok, F. (1995). *Commagene Nemrut*. İstanbul: *A* Turizm.
Demir, Ö. (1996). *Kapadokya, Medeniyetin Beşiği*. Nevşehir: Demir Color 7th Ed.
Demirulus, H. (2006). Journey to Fascinating Atmosphere of Nemrut Crater. *Ekoloji, Nature, Environment and Culture Magazine* İssue 12.
Dorner, R.&Friedrich, K.(1981). '*Kommagene*', Germany.
Goel, B T. (1996). *The Hierothesion of Antiochus I of Commagene* Volume: 2 Wiona Lake Indiana, Eisenbrauns.
Gökovalı,S. (1976). *Commagene,Arsameia, Nemroud Dagh*. İzmir: Arkeoloji Sanat.
Gökovalı, S. (1988). *Commagene and Nemru't*. İzmir: Arkeoloji Sanat.
Gray, M. (2007). '*Sacred Earth' Places of Peace and Power*. USA: Sterling.
Hamdi, O.& Osgan, E. (1987). *Le Tumulus de Nemroud-Dagh*. İstanbul: Arkeoloji ve Sanat.
Hudman, L.& Jackson, R. (2003). *Geography of Travel & Tourism*. NY: Thomson Delmar Learning. 4. Ed.
Önen, Ü. (1987). *Kommagene mit dem Götteberg Nemrut*, İzmir.
Sarıkaya, H E. (1982). *Türkiye Gizemleri*. İstanbul:Bilim Araştırma Merkezi.
USA Turkish Times. (2008). '*Mt. Nemrut: Eighth wonder of the ancient world.'*
Turkish News Agency, (2003). *Turkey 2003*. Ankara:Directorate General of Press & Information of the Prime Ministry.
http://www.sxc.hu/browse.phtml?f=search&txt=nemrut&w=1&x=21&y=11 8-04-2008.
http://web.deu.edu.tr/atiksu/ana37/ana37.html Nemrut Dagi ve Diablo II 8-04-2008.
http://en.wikipedia.org/wiki/Mount_Nemrut 8-04-2008.
http://www.travel-images.com/turkey11.html 8-04-2008.

Mt. Nemrud Caldera

Malike Özsoy

Mt. Nemrut is the southernmost and youngest of the chain of volcanoes in eastern Anatolia. It is a stratovolcano, (a composite volcano that is tall, conical and composed of many layers of hardened lava, tephra, and volcanic ash) and began erupting during the quaternary and continued to be active until 1597 AD, (Aydar et al., 2003).(It should not be confused with the "other" Nemrut Mountain, (Commagene) an archaeological site in southern Turkey which is now famous for its large carved stone heads).

The Situation of Mt Nemrud on the Map of Turkey

The volcano is named after mythical King Nimrod who it is said ruled this area in ancient times.

Nemrut mountain is 3,050 m high and has an elliptic caldera with a diameter of about 7 by 8 km (http://en.wikipedia.org/wiki/Nemrut *(volcano)*. It has a lake occupying the western third and numerous lava flows and domes (plus a couple more tiny lakes) covering the rest. The high point on the north rim of the

M. Özsoy (✉)
Architecture Department Yasar University, 35500 Bornova-İZMİR
e-mail: malike.ozsoy@yasar.edu.tr

N. Evelpidou et al. (eds.), *Natural Heritage from East to West*,
DOI 10.1007/978-3-642-01577-9_40,

caldera is about 2,300 ft (700 m) above the surface of the main crater, Lake Nemrut. *(*http://www.trekearth.com/gallery/Middle_East/Turkey/East_Anatolia/Bitlis/Tatvan/photo537121.htm*)*

Nemrut Volcano

Nemrut Volcano

As a result of the volcanic eruptions of Nemrut, the single Van-Mus river basin was divided into two separate basins. The eruption of Nemrut volcano also led to the formation of Lake Van, the largest lake in Turkey.

Bird's eye view of Nemrut Volcano located on the west side of Lake Van

Map of Nemrut Volcano and Lake Van

Nemrut is the southernmost volcano of the volcanic chain running from the Mt. Ararat in the easternmost part of Anatolia to the southwest. It is the only volcano in this volcanic chain which has records of volcanic activities in its history. (http://www.ersdac.or.jp/todayData/071/pict_e.html' *2007/04/02).*

It is located to the west of Lake Van. There are three lakes in the caldera. The western part of the caldera contains a large coldwater crater lake about 155 m deep. One lake, named Ilı göl, is hydrothermal. Its temperature reaches 60°C, providing evidence of continuing volcanic activity. (http://en.wikipedia.org/wiki/Nemrut_(volcano). Most of the lava flows were ryholite topped with obsidian. Much of this obsidian is beautiful rainbow obsidian which shimmers different colours in bright light.

A sample from obsidian

The Nemrut caldera has a climate peculiar to it; in other words, it has developed a local climate. Furthermore, the caldera is covered by vegetation, also peculiar to it, containing endemic species. Almost a natural monument, the volcano also accommodates wild animals (Demirulus, 2006). These include birds, for whom it is a paradise. With its unique bird species this place is a heaven for nature lovers.

Rock Sparrow

Finches, Linnet

Mountain Rock Thrush

Black Eard Wheatear

Starling Golden Eagle

Summer (June–September) is the best season for expeditions in Mt. Nemrut. Hikers who climb to the crater and summit from the southeast or eastern face of the mountain are rewarded with wonderful views of Lake Van and views of Nemrut Crater Lake.

The Nemrut Volcanic Caldera is the main refuge of some still-surviving glacial plants. Eastern sections of the Northern Anatolian Mountains, and the high mountains which are found in the inner part of the Anatolia, were occupied by the glaciers. Under the cold somewhat dry climatic condition, the Euro Siberian plants included Scotch pine (Pinus sylvestris). Birch (Betula) were widespread below the glaciated areas and migrated as far as western Anatolia and the southern high part of the Taurus mountains. During the present climatic condition these trees migrated towards the upper parts of the mountains, especially north facing slopes. The birch communities are found in the inner Nemrut volcanic caldera, (Atalay, 2006).

Crater Lake Crater Lake

Crater Lake

Crater Lake

Crater Lake

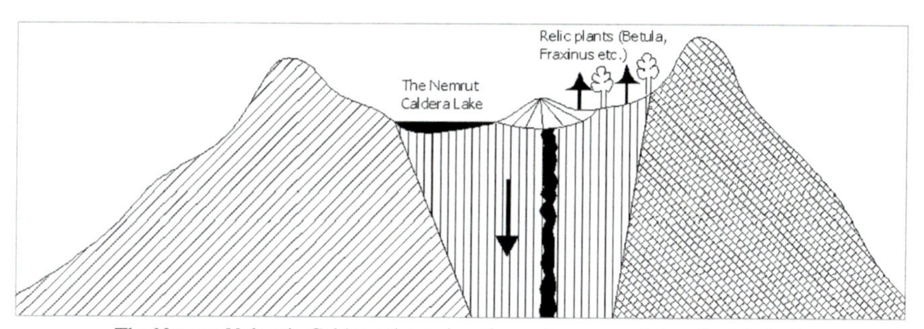

The Nemrut Volcanic Caldera: the main refuge of some surviving glacial plants

Nemrut Mountain also has very attractive vegetation features. The endemic plants such as Oak (*Quercus sp*), Birch (*Betula sp*) are types of tree species that grow in that area. The inside of the crater is also very important, since endemic plant species grow there. The plant "Sarmustahtayanı" (a kind of Betula) which takes its discoverer's name, only exists in Nemrut Caldera and nowhere else in the world. Moreover, other endemic species such as Creeping buttercup (*Ranunculus*) are unique to that area. Other natural species belonging to the Nemrut Caldera are *orchis umbrasa, orchis caucasia, orchis punetorum and epipactic latifolia.* (http://www.guroymak.gov.tr/kultur.php).

Birch

Scotch Pine

Poplar

Oak

Creeping buttercup

Orchid

Helleborine

Helleborine

References

Aydar, E., Gourgaud, A., Ulusoy, I., Digonnet, F., Labazuy P., Sen E., Bayhan, H., Kurttaş, T. & Tolluoğlu A.U. (2003). Morphological analysis of active Mount Nemrut stratovolcano, eastern Turkey: evidences and possible impact areas of future eruption. *Journal of Volcanology and Geothermal Research*. Vol.123, Issues 3. pp 301–312

Atalay, İ. (2006). The effects of mountainous areas on biodiversity: A case study from the northern Anatolian mountains and the Taurus mountains. *Grazer schriften der geographie und Raumforschung*. band 41 pp 17–26

Demirulus, H. (2006). Journey to Fascinating Atmosphere of Nemrut Crater. *Ekoloji Magazine*. October–December 2006: Issue 12

Turkish News Agency, (2003). *Turkey 2003*. Ankara: Directorate General of Press & Information of the Prime Ministry. http://www.trekearth.com/gallery/Middle_East/Turkey/East_Anatolia/Bitlis/Tatvan/photo537121.htm)

http://bullard.esc.cam.ac.uk/~maclenna/john_work_files/hugh_obsidian_615.jpg

http://www.crisafulli.com/images/animals/golden_eagle.jpg 2008/04/02

http://www.ersdac.or.jp/todayData/071/pict_e.htm 2008/04/02

http://www.guroymak.gov.tr/kultur.php2008/04/02

http://www.pbase.com/osmantanidik/Nemrut_bitlis 2008/04/12

http://www.resimlerim.org/NEMRUT-DA%C4%9EI-resimleri-5.aspx 2008/04/12

http://www.seaes.manchester.ac.uk/documents/newsletters/200609newsletter.pdf. 2008/04/12

http://www.skimountaineer.com/ROF/ 2008/04/02 ZX

http://www.sxc.hu/browse.phtml? f=search&txt=nemrut&w=1&x=21&y=1 2008/04/12

http://www.trekearth.com/gallery/Middle_East/Turkey/East_Anatolia/Bitlis/Tatvan/photo537121.htm) 2008/04/12

http://en.wikipedia.org/wiki/Nemrut_(volcano). 2008/04/12

Oludeniz Lagoon – Fethiye

Gokce Ozdemir and Osman Culha

Fethiye, ancient Telmessos, is the most important city of ancient Lycia; its exact date of establishment cannot be ascertained as yet (Aygen, 1988). However, the double "ss" in the last syllable of the name suggests that it was a local name (Aksit, 1984). The bay, which is well protected by an island, was especially important to navigation in ancient times and the city's location in a 30 km. square fertile plain, bordered by Caria in the north, was of equal importance (Önen, 1984). Fethiye is considered to be one of the most stunning tourist destinations of Turkey and Ölüdeniz is a well-known place with its outstanding panoramic view and natural resources.

The Location of Dead Sea (Ölüdeniz) - Fethiye on the Map of Turkey
Source: Ministry of Culture and Tourism

Fethiye's inner bay is an excellent natural harbor, protected from storms by the Knight's Island. The much larger outer bay has 11 more islands (Brosnahan and Yale, 1996). The island originally named Makri Vecchia, was later given the name

G. Ozdemir (✉)
Department of Tourism and Hotel Management, Yasar University Kazim Dirik M., 364 S, Bornova, Izmir, Turkey
e-mail: gokce.ozdemir@yasar.edu.tr

N. Evelpidou et al. (eds.), *Natural Heritage from East to West*,
DOI 10.1007/978-3-642-01577-9_41, © Springer-Verlag Berlin Heidelberg 2010

"Isola Longa" by the Venetians and Genoans (Aygen, 1988). The medieval battlements were reinforced by the Knights of Rhodes, who tried to gain control of the region from the Knight's island, in the bay of Fethiye, where they also built a fortress and controlled the town of Fethiye (Aksit, 1984).

Beaches in Ölüdeniz were formed in three ways. The first, the way in which present beaches have been created, occurs when pebbles brought down by rivers are shaped by the waves of sea. The second is the shaping by waves of screes falling from terrestrial slopes. Finally, Ölüdeniz lagoon was formed as the result of the collapse of shore caves (Öztürk et al., 2003).

Kayaköy, called Levissi for much of its history, is a town of stone houses near Hisarönü; it was deserted by its mostly Ottoman Greek inhabitants after the First World War and the Turkish War of Independence. (Brosnahan and Yale, 1996). Consisting of 3500 houses with completely Greek inhabitants, the village was abandoned in 1922 when all the occupants left the area, burning and destroying all that was left behind. After this Turks migrated from the Balkan Peninsula, settled here and constructed a new town near the old city (Aygen, 1988).

In addition to the Knight's Island, attractive places within Fethiye are Kayaköy, and particularly Ölüdeniz (Dead Sea) with its beautiful Butterfly Valley. Ölüdeniz is a sheltered lagoon that has splendid natural beauty and is totally hidden from the open sea. The view of Ölüdeniz up from the hills provides an incomparable expression of natural magnificence with a composition of green (a pine forest that surrounds the sheltered lagoon) and blue (an open sea that is connected to Ölüdeniz

Oludeniz Lagoon (Dead Sea)
Source: EGENET Tourism Promotion Limited Company

Paragliding on the Lagoon - Fethiye
Source: EGENET Tourism Promotion Limited Company

with a 2 km long sandy beach). Kumburnu forest bordering the lagoon is a protected area which is also one of the national parks of Turkey. Additionally, Ölüdeniz has been a natural port for the yachts throughout history; now it is a fascinating sight-seeing area for both blue voyage tours and private yachts. Most of all, experiencing many shades of blue and green fascinates tourists from all over the world.

Natural Port Ölüdeniz - Fethiye
Source: EGENET Tourism Promotion Limited Company

The "Butterfly Valley", one of the most beautiful coves of Ölüdeniz, is located on an area of 10 ha at the bottom of a steep and deep valley. The valley is like an open-air museum of local fauna, displaying the life-span of 30 daytime and 40 night time species of butterflies and moths for nine months starting in the spring (Dinçer and Tor, 2005). The valley takes its name from the butterflies and moths such as the *Danaus chrysippus, Limenitis reducta, Utetheisa pulchella, Melitaea didyma, Papilio machaon* and many more. The "Jersey Tiger" moth (*Euplagia quadripunctaria*), which appears mostly in July and August (Ergül and Doğan, 2004), is the most rare of the lepidoptera in the valley. There are also 12 endemic and vulnerable plant species which have been isolated and protected in the "Butterfly Valley". *Campanula hagelia, Sideritis albiflora, Quercus aucheri, Stachys bombycina, Galium canum, Phlomis lycia*, and *Dianthus elegans* are some of the examples of endemic plant species. Examples of vulnerable plant species are *Brassica cretica, Euphorbia dendroides, Arum nickelii, Orchis sancta* L., and *Ophrys holoserica* (Senol and Yıldırım, 2004).

The Butterfly Valley - Fethiye
Source: EGENET Tourism Promotion Limited Company

Butterflies in the Butterfly Valley - Fethiye
Source: EGENET Tourism Promotion Limited Company

On the 8th of February 1995, the site was declared a first degree natural protected area and any type of construction has been prohibited. The valley is rocky and pine covered and is a bit hard for climbing but it gives you an astonishing feeling to see millions of butterflies covering the trees and rocks like a soft colorful scarf (Mugla Governorship, 2005).

References

Akşit, İ. 1984. Blue Journey to Lycia and Pamphylia. Istanbul: Akşit Culture and Tourism Publications.

Aygen, T. 1988. The Blue Paradise of Lycia. İstanbul: Dost Publications.

Brosnahan, T. & Yale, P. 1996. Turkey a Lonely Planet Travel Survival Kit. Australia: Lonely Planet Publications 5th edition.

Dinçer, N. & Tor, K. H. (Editor), Tor, K. H. (Drawings), Tor, K. H., Okutan, O., Akbaş, F., Tüzer, M. and Tekin A. (Photography), Duru, E. (Translation), Çaçaron, U. and Taşçı S. (Redaction). 2005 Municipality of Ölüdeniz – Fethiye. Ankara: Offset Printer.

Ergül, B. & Doğan, A. 2004. Tempo A'dan Z'ye Fethiye Muğla 4 Rehberi. Ankara: Doğan Burda Rizzolli Dergi Yayıncılık ve Pazarlama A.Ş.

Muğla Governorship. 2005. Muğla, Bodrum, Dalaman, Datça, Fethiye, Kavaklıdere, Köyceğiz, Marmaris, Milas, Ortaca, Ula, Yatağan. Istanbul: Seçil Offset.

Şenol, S.G. & Yıldırım, H. 2004. Ecotourism for Endemic and Vulnerable Species in The Butterfly Valley. Tüdav Eğitim Serisi No: 8, Istanbul.

Önen, Ü. 1984. Lycia Western Section of the Southern Anatolian Coast Antique Cities-History-Works of Art. Izmir.

Öztürk, B. Öztürk, H. & Görgün, M. 2003. Sea Ambient and Dead Sea. Tüdav Eğitim Serisi No:1, Istanbul.

Pamukkale (Hierapolis)

Gokce Ozdemir

Pamukkale, also called "Castle of Cotton", offers a highly picturesque and unique mix of natural beauty with its amazing limestone formations. Pamukkale (Hierapolis) is designated as a mix of cultural and natural heritage having outstanding universal value, by the UNESCO World Heritage Center. The ancient city of Hierapolis, regarded as a model for spas of the present day, represents the distinctiveness of the place. The ancient remains like the City Walls, Theater, St. Philip's Martyrium, Churches, Great Bath Complex, Temple of Apollo, and Necropolis are worth seeing in and around Hierapolis. It is striking that hot springs with therapeutic powers accompany these ruins so harmoniously. This composition of culture and nature creates a marvelous panorama with a unique visual experience attracting many visitors from all around the globe. It is wealth of history that makes Pamukkale a popular tourist destination in Turkey.

The Location of Pamukkale (Hierapolis) on the Map of Turkey
Source: Ministry of Culture and Tourism

Tectonic movements taking place in the fault depression of the Menderes river basin gave rise to the emergence of a number of very hot springs, and it is the water from one of these springs, with its large calcium carbonate (calcite) mineral content, that has created the natural wonder of travertines (http://www.turizm.

G. Ozdemir (✉)
Department of Tourism and Hotel Management, Yasar University Kazim Dirik M, 364 S, Bornova, Izmir, Turkey
e-mail: gokce.ozdemir@yasar.edu.tr

N. Evelpidou et al. (eds.), *Natural Heritage from East to West*,
DOI 10.1007/978-3-642-01577-9_42, © Springer-Verlag Berlin Heidelberg 2010

net/cities/hierapolis/). The term "travertine" comes from the Italian "travertine" or "tivertino", and is derived from the Latin *lapis tiburtinus*, "tiburtine rock"; Tibur was the name of Tivoli (where there are extensive deposits of the mineral) in Roman times (Pamukkale, 1992). The travertines are composed of a sequence of terraced hills that change color according to the angle of the sun's rays. These fantastic hills, which from afar stand out against the lush valley, seem like an enormous cotton field (Valdes, 1993). The terraced layers and steps in this interesting natural landscape have formed little by little for centuries and, now provide a fascinating scenic attraction for the visitors.

Limestone Formations
Source: Governance of Denizli

The medicinal value of this beneficial mineral water, rich in calcium, magnesium sulfate, bicarbonate and carbon dioxide was understood by every civilization. As they do today, people in ancient times sought cures for their diseases and tried to benefit from the healing opportunities offered by the thermal spring. As Toksöz (1989) stated, fascinating limestone formations have been built up by the streams that for thousands of years have flowed down from the slopes; and beside the source of the hot springs the old sanctuary formed the nucleus of Hierapolis, a city that played such an active role in both the Roman and Christian periods. In Hierapolis, there is a complex that includes the Sacred Pool, a colonnaded street, and a basilica church. The Sacred Pool is warmed by hot springs and littered with underwater fragments of ancient marble columns. During the Roman period, columned porticoes surrounded the pool; earthquakes toppled them into the water where they lie today. Possibly associated with the Temple of Apollo, the pool provides today's visitors a rare opportunity to swim with antiquities (http://www.sacred-destinations.com/turkey/hierapolis-pamukkale.htm).

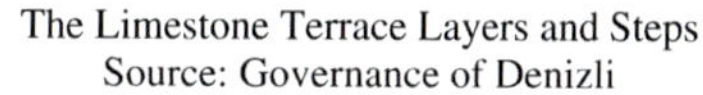

The Limestone Terrace Layers and Steps
Source: Governance of Denizli

Hot Springs and Thermal Pools in Hierapolis
Source: Governance of Denizli

In all periods, Pamukkale has had a reputation for the hot springs rich in calcium and suitable for thermal bathing. Hence, there are many legends about the hot springs and thermal pools. As Valdes (1993) pointed out, leaving the interesting myths aside, the waters of Pamukkale gush at 35 degrees centigrade; they contain considerable amounts of salts, lime, carbon dioxide and minerals so they are famous for their healing and therapeutic properties. When the thermal waters emerge from the springs and the pressure is suddenly removed the carbon dioxide with which the water is heavily charged tends to vaporize and calcium carbonate is precipitated, gradually forming the fascinating travertine formations (Yavi and Yavi, 1998).

Travertine Terraces
Source: Governance of Denizli

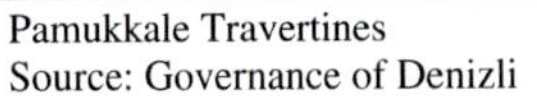
Pamukkale Travertines
Source: Governance of Denizli

Hierapolis
Source: Governance of Denizli

The chemical composition of the underground water, in the Pamukkale travertine area, having the characteristic of calcium bicarbonate ($CaHCO_3$) and a high level of partial carbon dioxide pressure, cause the formation of travertine under atmospheric conditions. The amount of calcium carbonate in the natural water is directly related to pH values and the main component controlling the pH values of this water is partial carbon dioxide pressure (pCO_2) (Açıkel, 2006: 11). Loss of carbon dioxide (CO_2) gives rise to increasing the saturation level of water for calcite, consequently the process of precipitation of travertines starts. The faster the loss of carbon dioxide, the more the travertine precipitates. The rate at which carbon dioxide comes out of solution in the water depends on the water's speed of flow. The change of water from regular flow into gush flow on the terraces, and the spreading of the water out over the terraces, causes the loss of CO_2 gas in the water much more rapidly. However, in the places where the water is deep and the flow regular, for instance, in the canals that have a gentle slope, the loss of CO_2 is much slower. The loss of CO_2, during the flow of water from the springs to the travertine terraces, reaches its highest level while passing from canals to the travertine area and at the points where small waterfalls are occurring at the terraces (Ekmekçi et al., 1995; Açıkel, 2006).

Today this site is a most striking and unusual marvel of nature. From the plain one can see impressive petrified waterfalls, which are what the sparkling white terraces, where the waters collect, actually resemble. The image is fascinating and enchanting at the same time (Valdes, 1993).

References

Açıkel, Ş. (2006). Determination of Solute Transport Parameters of Pamukkale (Denizli) Travertines and Evaluation With Regard to Whitening Processes. Unpublished Dissertion, Hacettepe University, Ankara.

Ekmekçi, M., Günay, G., Şimşek, Ş., Yeşertener, C., Elhatip, H. & Dilsiz, C. (1995). CO2 Outgassing Precipitation Kinetics Relations In Travertine Depositing Hot Waters Of Pamukkale, An Earth Sciences Journal, 17: 101–113.

Denizli (Turkey) Governorship, Ministry of Tourism and, UNESCO. (1992). Pamukkale (Hierapolis) Preservation and Development Plan and the International Workshop on Pamukkale 30 June–July 1991. Ankara: Dönmez Ofset.

Pamukkale, Retrieved from the World Wide Web March 25, 2008: http://www.turizm.net/cities/hierapolis/.

Sacred Destination Travel Guide Explore the Spiritual Heritage of the World. Hierapolis (Pamukkale). Retrieved March 25, 2008 from the World Wide Web: http://www.sacred-destinations.com/turkey/hierapolis-pamukkale.htm.

Toksöz, C. (1989). Pamukkale and Hieropolis. Istanbul: Toksöz Publishes.

Valdes, G. (1993). Pamukkale (Hierapolis-Iaodices Aphredisias).

Yavi, E. & Yavi, N. Y. (1998). Türkiye'nin Parlayan Yıldızı Denizli. Denizli: T.C. Denizli Governorship.

Index

N. Evelpidou et al. (eds.), *Natural Heritage from East to West*,
DOI 10.1007/978-3-642-01577-9, © Springer-Verlag Berlin Heidelberg 2010

D